FREE-LIVING
FRESHWATER
PROTOZOA
A COLOUR GUIDE

Free-Living Freshwater Protozoa

A Colour Guide

D.J. Patterson BSc PhD DSc
Professor of Biology, University of Sydney

Drawings by Stuart Hedley BSc

John Wiley & Sons
NEW YORK TORONTO

MANSON
PUBLISHING

UNSW PRESS
SYDNEY

...ing Ltd

...may be reproduced, stored in a retrieval
...ans without the written
...ance with the provisions of the Copyright
...ny licence permitting
limited copying issued by the Copyright Licensing Agency, 33–34 Alfred Place,
London WC1E 7DP.

Any person who does any unauthorised act in relation to this publication may be
liable to criminal prosecution and civil claims for damages.

A CIP catalogue record for this book is available from the British Library.

Published in **North & South America** by
John Wiley & Sons Inc, 605 Third Avenue, New York, NY 10158–0012
ISBN 0–470–23567–5

Library of Congress Cataloging-in-Publication Data
Patterson, David J
      Free-living freshwater protozoa: a colour guide/D.J. Patterson:
      illustrated by Stuart Hedley.
      223 p.   19.4x26.1 cm.
      Includes bibliographical references and index.
      ISBN 0–470–23567–5 (hc)
         1. Freshwater protozoa – Classification.   2. Freshwater protozoa – Identification.
3. Freshwater protozoa – Atlases.   I. Title.
QL366.5.P37   1996
593. 1'0916'9–dc20
                                                                96–38713
                                                                CIP

Published in **Australia and New Zealand** by
University of New South Wales Press Ltd, Sydney, NSW 2052.
ISBN 0–86840–327–X

National Library of Australia Cataloguing-in-Publication Data
Patterson, David J
Free-living freshwater protozoa: a colour guide.
Bibliography.
Includes index.
ISBN 0–86840–327–X

1. Freshwater protozoa – Classification. 2. Freshwater protozoa –
Identification.   I. Title
593.109169

Printed and bound in England

Contents

Daytona

Preface

Protozoa are important consumers in many aquatic ecosystems including some of economic importance, such as waste water-treatment plants. The biology of such habitats cannot be properly described unless the protozoa are taken into account. Protozoan communities respond rapidly to changing physical and chemical characteristics, and can be useful indicators of changes in natural communities. In spite of these facts, the inclusion of a protozoan perspective in studies of the biology of aquatic habitats, especially by non-protozoologists, is relatively rare. The purpose of this Guide is to make free-living protozoa a little more accessible to students and professionals who need to study protozoa in the course of their work.

The emphasis of the Guide is on those organisms that are most likely to be encountered in freshwater habitats. It is assumed that most readers wanting to develop a basic knowledge of free-living protozoa will not use specialist staining or preservation techniques, rather they will observe living organisms. The identification of most genera is relatively straightforward, but that of most species is more difficult, requiring specialist techniques. Those who wish to work at the species level will find reference to suitable literature in the Bibliography.

Acknowledgements

A variety of people have provided advice or other forms of support. I would like to thank Hilda Canter, Marina Christopher, Tore Lindholm, Craig Sandgren and Helge Thomsen for photographic material; Klaus Hausmann for teaching me the techniques of photomicrography; Willi Foissner, Jacob Larsen, Øjvind Moestrup, Colin Ogden, Ferry Siemensma, Michael Sleigh, Naja Vørs and Michael Zölffel for confirming and/or correcting identifications, or for commenting on the key and the text, and E.G. Bellinger, Hilary Kelle and Richard Foot (Wessex Water) for advice on water treatment. The assistance of Claire Wyatt and Kevin Williams is also deeply appreciated. The University of Bristol Publication Fund provided financial assistance. Finally, I would like to thank Stuart Hedley, without whose interest and skills this Guide would not have appeared.

D.J.P.

Introduction

Protozoa may be found in almost every aquatic habitat, from cesspit to mountain stream, from garden birdbath to the Amazon. Natural communities typically contain dozens of species, and this diversity is retained when collections are made in large jars and returned to the laboratory. The richness is expressed as a spectacular array of body forms, reflecting the wide range of niches occupied. The number of species, the number of individuals within each species, and the types of species can all provide valuable insights into the nature of the habitat from which a sample was taken. For these reasons, protozoa can be a convenient source of material to illustrate biological principles.

In recent years it has become clear that, despite their small size, the contribution of protozoa to the metabolism of aquatic and terrestrial ecosystems can be very substantial. Protozoa occupy a significant, sometimes dominant, position among the consumers within a community. Their importance is closely linked with their use of bacteria as a source of food. Progress in understanding the role of particular protozoa and the nature of the transactions that occur among members of the microbial communities has been held back because there is little familiarity with the organisms.

Protozoan communities are very dynamic structures, with numbers of cells changing rapidly by cell division, encystment or excystment. The structure of the protozoan community quickly responds to changing physical and chemical characteristics of the environment, suggesting a potential use of the diversity of protozoa and the occurrence of particular species as indicators of changes in ecosystems. However, such suggestions should be followed with caution as there are major difficulties involved in finding the right way to sample the habitats occupied by protozoa, and in accurately identifying species.

In order to understand or to use the protozoan community (particularly as an indicator of change) in teaching or research, it is necessary to be able to identify individual protozoa. Developing familiarity with the diversity of any group of organisms can be a daunting task. In the case of protozoa, there are few English-language books offering an authoritative understanding of the group. Many books claiming to be guides to the protozoa (see Bibliography, for a list of such books) often require knowledge of special preservation and/or staining techniques, or they rely on codified drawings of organisms. The latter may be suitable for specialists, who understand the way in which the information has been coded, but such drawings add to the hurdles faced by beginners. This Guide relies heavily on photographs because they show protozoa as they would appear to an observer looking down a microscope, and make the learning and recognition processes a little more exact.

This Guide deals only with protozoa from freshwater sites, and the organisms illustrated have been obtained from ditches, small ponds, puddles, lakes, aquaria and water-treatment plants. The common and accessible organisms are emphasized in preference to rarer organisms or organisms less likely to be collected (e.g. from open lakes or from anoxic habitats). As the book is not comprehensive, readers will find that some organisms are not included here. For these, and for identification beyond genus, the user should make use of the specialist literature cited in the Key and the Bibliography.

The terms 'free-living', 'protozoa' and 'freshwater' have been rather freely interpreted: included are a few organisms that may be found in soils, mosses and low-salinity brackish water habitats. There is considerable species overlap between communities from these habitats and from more usual freshwater sites. Indeed, some species can be found in both marine and freshwater sites. Soil protozoa are of great importance in cycling nutrients, and should not be ignored. Also included are a couple of protozoa that are found living attached to other organisms. Ectosymbionts rarely harm their hosts, and are best regarded as free-living species exploiting their hosts for attachment. Their location usually secures a better supply of food.

What are protozoa?

Protozoa cannot be easily defined because they are diverse and are often only distantly related to each other. They are unicellular eukaryotes. Together with the unicellular algae and the slime moulds, they make up the protists. Protozoa have usually been distinguished from algae because they obtain energy and nutrients by heterotrophy, that is, by taking in complex organic molecules, either in soluble form (osmotrophy) or as particles such as bacteria, detritus or other protists (phagotrophy). Protozoa ('first animals') get their name because they employ the same type of feeding strategy as animals. Heterotrophy contrasts with photosynthesis, the use of radiant energy (sunlight), as a source of energy for metabolism (as in algae and plants). In unicellular organisms, these two nutritional strategies are not mutually exclusive (as they are in multicellular plants and animals). Indeed, quite a large number of flagellates are mixotrophic and can use both types of nutrition; many heterotrophic protozoa harbour photosynthetically active endosymbionts.

Protozoa include amoeboid, flagellated and ciliated organisms that are capable of heterotrophic nutrition, whether or not they also contain chloroplasts (see p.19, for the composition of protozoa). Some heterotrophic protists evolved before the ability to photosynthesize was acquired, but others evolved from algal protists by loss of their chloroplasts. Not only has the protozoan state been achieved independently in different lines of evolution, but these organisms cover an immense area of evolutionary territory; measured in molecular terms, two protozoa may have less in common than do a plant and an animal. Furthermore, not all protozoa are equally equipped to deal with the demands of living. Having appeared over the period during which the eukaryotic cell was being assembled, some have relatively few organelles at their disposal, whereas others have been very inventive in the development and application of organelles. One should therefore be very wary of making generalizations about this diverse group.

Distinguishing protozoa from other microbes

The microscopic community includes bacteria and blue-green algae (both are prokaryotic, having no nuclei or other discrete cellular organelles), algae (both motile and non-motile, unicellular and multicellular), slime moulds, and some small (multicellular) metazoa (especially rotifers, gastrotrichs, copepods, flatworms and nematodes). Typical representatives have been illustrated (Figs 4–19) to aid in distinguishing protozoa from non-protozoa.

In principle, distinguishing protozoa from metazoa is simple: protozoa are single-celled; metazoa are comprised of many cells. In practice, it is extremely difficult to see the boundaries of cells within microscopic metazoa, and other features may have to be relied upon for positive identification: the exoskeleton and jointed appendages of small arthropods (Fig. 231) allow them to be easily recognized; rotifers have a distinctive mastax (Fig. 15) behind the mouth, anterior cilia and a forked 'tail'; and the round, smooth surface of most nematodes (Fig. 18), together with their stiffness and their serpentine motion, makes them easy to recognize. Only a few gastrotrichs (Fig. 16) and tardigrades (Fig. 19) are likely to be found, and they are quickly learned.

Equally simple on paper is the distinction between protozoa and prokaryotes. The latter have no internal organelles, and are usually very much smaller than protozoa, with one dimension restricted to about 1 μm. However, some protozoa are very tiny, and some bacteria are rather large and may have various inclusions, so identification is not always straightforward. Most bacteria have a simple shape (spherical, sausage-shaped, helical), and most, but not all, are rigid. Many are capable of swimming or gliding, but none have the lashing cilia or flagella of eukaryotic single cells, nor are they able to form pseudopodia.

Protozoa may be amoeboid, flagellated or ciliated. There is no clear boundary between flagellated protozoa and flagellated algae. If definitions of algae and protozoa based on their respective nutritional strategies are accepted, then some organisms are both algae and protozoa. Some protists that rely exclusively on heterotrophy (i.e. are protozoa) may be closely related to some species that rely exclusively on autotrophy (i.e. are algae). Exceptionally, autotrophic and heterotrophic species may be so closely related that they are placed in the same genus. As no clear boundary can be drawn here, this Guide includes some algal cells that may be a source of confusion, or that are closely related to protozoa.

Just as there is an unclear boundary between algae and colourless flagellates, so there are problems in distinguishing between slime moulds and amoebae. Slime moulds are amoeboid organisms with two stages in their life cycle that are not encountered in conventional amoebae. They can form large amoeboid masses, and they may produce a mass of spores (cysts) that is lifted away from the substrate on a stalk, allowing the spores to be released into air or water currents to aid in the dispersal of the organism. The large amoeboid stages (plasmodia) are rarely observed, except when special growth media are employed, and have not been included here. Slime moulds may have unicellular (amoeboid or flagellated (Fig. 20)) stages in their life cycle, which may be misidentified as protozoa. Some hyphal fungi also produce flagellated stages (zoospores) which may be mistaken for protozoa.

Equipment

For effective protozoological work, the following equipment is essential: a compound microscope, 2.6 × 7.6 cm glass slides with coverslips, glass dropping (Pasteur) pipettes and teats, small glass Petri dishes, collecting jars, soft tissues and filter paper (blotting paper). Ideally, the following equipment should also be on hand: a dissecting (low-power or binocular) microscope, an alcohol or bunsen burner, lens tissue, a measuring eyepiece and micrometer slide, a filter apparatus (filter funnel/coffee filter), plankton net (less than 20µm mesh), a can of compressed air with a nozzle, a collection of bottles and jars, solid watch glasses, barley, wheat and/or rice grains, agar powder, autoclave (pressure cooker), photographic facilities and a centrifuge.

Microscopes

The choice of microscope is important in the study of protozoa, in that a good choice will allow you to see more, and to see it with ease. Microscopes with built-in illuminators and binocular eyepieces are more convenient than those with separate light sources, and protozoologists also benefit greatly from phase contrast optics. The condenser should be equipped with an iris. An option for photography, such as a trinocular head with a vertical tube to which a camera can be attached, is desirable.

The components requiring the most critical consideration are objectives: the best affordable should be used. Normally several objectives will be needed, and the magnifications should range between x10 and x100.

Magnifying power is less important than resolving power, that is, it is far better to see details clearly than to have them appear large but blurred. If finances are restricted, it is preferable to buy a smaller range (minimum of two, about x10 and x40) of good-quality objectives than several of poor-quality. Higher-power objectives are usually of the oil-immersion type. Ideally, if the microscope is capable of phase contrast microscopy, phase objectives should be bought in preference to those of the bright-field (normal) variety. The eyepieces should have a magnification of between x8 and x12.

A dissecting microscope is a lower-power microscope with greater depth of field than a compound microscope, a longer working distance (between the lenses and the specimen), and usually with stereoscopic vision. It is ideally suited to hunting around a sample before material or organisms are selected for observation or culture. Lighting on such microscopes may either be from above (top lighting) or transmitted through the specimen. The latter is desirable when examining protozoa, with the light source removed as far as possible from the organisms to increase contrast and to reduce the risk of heating the specimen. Good dissecting microscopes have built-in illuminators.

Compact field microscopes are also available.

Basic care of microscopes

Microscopes are expensive and delicate, and the glass surfaces are most vulnerable to dirt and damage. Both cause reduction in image quality. For example, mascara-laden eyelashes can damage the surfaces of eyepieces! However, a properly maintained microscope can last for decades.

As far as possible, keep dust off microscopes by protecting them with a plastic bag or cover. Do not leave any open tubes uncovered (e.g. the eyepiece tubes), as dust will get inside the microscope. Avoid sudden changes in temperature, since this can lead to condensation inside the microscope. Do not place a microscope where it can be splashed with water or other chemicals. Salt water should be removed quickly if it gets onto the microscope. As a general rule, microscopes should be kept in a dry atmosphere and at an even temperature.

If possible, avoid touching glass surfaces with any material. Most dirt will be in the form of dust, and it is best cleaned off using compressed air from canisters (such as can be bought at photographic agencies). If surfaces have to be cleaned by wiping them (e.g. to remove immersion oil), avoid using any materials that may contain grit, such as cheap paper tissues. Special lens tissues are available (usually from photographers and opticians) for cleaning optical glass surfaces, but clean, soft cotton is also very good.

Lubrication of moving parts (stage movement, objective turret, focusing mechanism, iris diaphragms) is best left to experts.

Basic microscopy

Familiarize yourself with the principle components of the microscope. These include the light source, condenser, stage, objectives, and eyepieces.

● **The light source:** almost all modern microscopes have built-in illuminators, typically equipped with a diffuser to give even illumination, and a regulator to control the level of illumination. There is no 'perfect' level of illumination: light intensity should be adjusted for personal convenience. More intense lights tend to heat the specimen being observed, and will lead to physiological distress and morphological distortion of cells. Minimal illumination, best achieved by working in a dimly lit or darkened room, is desirable.

Some microscopes have meters that indicate the relative intensity of illumination. In some, the upper extreme is marked (usually in red), and if the intensity remains at this level for a long time, it will shorten the life span of the bulb. Some meters also have a marked zone in the middle of the range, within which the best colour balance in photographs will be achieved.

● **The condenser** is a lens system that focuses light onto the specimen. The condenser can be moved up and down relative to the stage, and on some microscopes it may also be possible to move it to the left and the right, and backwards and forwards. It may contain an iris, and may have removable or optional components for different contrast enhancement techniques.

● **The stage** usually has a clip that is pressed against the end of a slide to hold it in place. Additional clips that press onto the top of a slide are entirely unnecessary.

● **The objectives** are usually located on a rotating turret, and will click into place. If all the objectives have been bought from the same manufacturer, they should all focus at the same level, eliminating the need to change the focus as you switch from one objective to another, and helping to prevent accidental damage to specimens, or to the objectives themselves.

The highest-power objectives are usually of the oil-immersion type. A drop of special immersion oil, which should be obtained from the microscope manufacturer, is placed on the coverslip above the specimen, and the objective is then rotated into place so that it touches the oil.

To set a microscope up for basic bright-field (no contrast enhancement) microscopy, the steps are as follows:

● First select a low-power objective (x10 or lower). Place the slide (with coverslip) on the stage, switch on the lamp, check that all irises (lamp housing and condenser) are wide open, and focus on the specimen or on the edge of the slide or the coverslip.

● If there is an iris in the lamp housing, close it down; if not, place an object with a distinct edge (e.g. a piece of paper) on the glass surface of the lamp housing that is nearest the specimen. Looking down the microscope, adjust the condenser until the edge of the iris or of the piece of paper is in focus. The condenser is now focused to project light onto the specimen.

● If there is a lamp iris, make sure that its image is in the middle of the field of view. If this is not the case, then the condenser is projecting the light to one side of the objective rather than along its optical axis. Check that the objective is screwed in tightly and that it is clicked into its proper position. If the light is still directed to one side, the condenser may be incorrectly fitted, or you may have to adjust its side-to-side or to-and-fro position in order to align it along the optical axis of the objective. There are usually two knobs or screws for this purpose. Having centred the condenser, open the lamp iris.

The above steps need only be repeated at the outset of each session. The following steps should be carried out every time the objective is changed:

● Remove one eyepiece, and, looking straight down the tube, close the condenser iris if there is one. Open the iris until any change in its position neither enlarges the area being illuminated nor increases the amount of illumination. Having closed the iris slightly, replace the eyepiece. The microscope is now ready for use.

The illumination achieved by setting up the microscope in this way is called bright-field microscopy. Specimens with colour and great inherent contrast can be seen clearly, but most protozoa will be almost impossible to see. Consequently, some form of contrast enhancement will be required.

Contrast enhancement

Contrary to the recommendations of many books on microscopy, resolving power is less important to protozoologists than is visibility. The lack of optical contrast in many protozoa means that very little can be seen using normal bright-field microscopy. The photographs of *Paramecium bursaria* (Figs 349–358) illustrate some of the techniques that may be employed to enhance contrast. Special accessories are required for most of these.

The simplest way of enhancing contrast is to close the iris in the condenser, or to lower the condenser so that it is below its optimal position (compare Figs 349 & 350). Dark-ground effects (Fig. 353) can also be achieved by adjusting the lighting so that light is directed through the object, but passes to one side of the objective. Both techniques make specimens visible, but they can only reveal a limited amount of detail in an object.

Phase contrast facilities are widely available. Special objectives and a condenser are required. If starting from scratch, it is probably more economical to buy phase facilities at the outset. Phase contrast is a relatively cost-effective way of getting good, high-contrast images. Nomarski (differential interference) contrast also requires special accessories; these are relatively expensive, but the resulting images (Fig. 352) have great clarity as well as good contrast.

Microscopical examination

Normally, preparations are made on glass microscope slides; a coverslip should always be used, as it protects the objectives from contamination and improves image quality.

2.6 x 7.6 cm reusable glass slides are widely employed. They should be cleaned and polished with tissue before use. Coverslips are available in various sizes and thicknesses, ranging from No.0 to No.2, which are thin and thick respectively. The author recommends 22 x 22 mm (square) or 32 x 22 mm No.1 coverslips. Like slides, coverslips should be cleaned before use, as small glass fragments, or greasy films reduce image quality. Coverslips are cleaned by carefully drawing them between folds of tissue held between the thumb and forefinger.

Heat and oxygen-depletion can cause cells to become moribund. The lamp of a microscope tends to warm specimens, and cells may only remain healthy for a few minutes. Bringing samples from bodies of water into a laboratory can involve a 10–20°C temperature change, which is enough to cause extensive physiological distress. Samples taken from organically enriched sites (e.g. sewage treatment plants) and placed under a coverslip will rapidly use up the available oxygen, and the community structure will begin to change within a few minutes. Thus, rapid processing is usually desirable if you wish to observe healthy cells behaving normally.

It is usually more convenient to add only a small drop of the sample to the slide. If it is possible to move the coverslip around freely, there is too much fluid, and protozoa will move not only in the lateral plane but also in a vertical plane, making careful observation almost impossible.

The movements of protozoa often cause problems. Usually, active motility is a sign of distress. Typical causes might be pressure from a coverslip, overheating, or depletion of oxygen. The cells move until they find a more favourable site. The use of minimal illumination or gently blowing on a preparation as you observe it often 'calms' protozoa. Filters that remove heat can be obtained from microscope suppliers. If these devices do not work, various narcotizing agents or viscous slowing agents can be used. Narcotizing agents include the solutions of heavy metals, such as nickel or copper chloride (used at a concentration of 5–10 mMol/l), while methyl cellulose can be used to increase viscosity. Iodine or formalin will kill protozoa. All of these treatments may cause distortion of one sort or another, and, as one of the great pleasures of watching protozoa is to see them behaving naturally, all can be regarded as unsatisfactory

An alternative means of immobilizing active organisms is to use a small piece of tissue paper to draw excess fluid from under the coverslip. The coverslip is pulled down towards the slide, and protozoa can then be trapped. Such samples can be observed for about five minutes before cells become distorted. A small pipette and some fluid should be kept handy, as it may be necessary to add a small drop of fluid to the edge of the coverslip to release the cell from terminal compression.

A few protozoa may go unnoticed because of their inactivity. Amoebae, in particular, require a few minutes to recover from the trauma of being placed on a slide. Other organisms may be located in detritus and will not become visible until they have been given sufficient time to disperse from it.

Complete beginners are advised to work with material that is known to contain many protozoa, for example, natural samples that have been checked using a binocular microscope, cultures from biological suppliers, sludge samples from treated sewage, coverslips left for three days on mud collected when there was an orange or green patch on the surface, or water with a soup-like consistency. With samples maintained in bottles in the laboratory, the fluid in the middle section of the bottle will have relatively few organisms; most protozoa will be found near the sediment or associated with the surface film. This can be sampled by placing the flat side of a coverslip against it.

Using one of the methods for enhanced contrast, and making sure that the microscope is focused on the sample (check the edge of the coverslip) at a low magnification (about x 10), scan the slide methodically to find protozoa.

In order to examine rare or specific types of protozoa, it may be necessary to soften the glass of a Pasteur pipette in a burner, remove it from the flame, and, with a smooth movement, draw it out to the thickness of a hair. This pipette can then be broken to give an aperture with a diameter 2–5 times greater than that of the cell, and can be used to pick up individual cells with the minimum amount of fluid while using a dissecting microscope. Protozoa collected individually or in small numbers have the peculiar ability to disappear after being added to a slide: they may be killed as they are pulled into the pipette, adhere to the inner surface of the pipette, be smashed as the coverslip is added, or move to the thin film of fluid around the outside of the coverslip. Care and regular examination of the cells throughout the procedure are advisable if it is important that a particular species is observed.

Large protozoa may be crushed by coverslips, and should be protected by creating a chamber on the slide. This is achieved by placing two shards of broken coverslips on either side of a drop of fluid and then laying another coverslip across them.

Recording protozoa

It is strongly recommended that protozoa should not merely be observed, but also recorded. The most simple and often the best way of recording protozoa in a sample is to make line drawings. This directs the eye to important features. The copying of drawings from books should be avoided, as many are inaccurate and often contain information in a coded form.

A picture of the organism should be built up, beginning with outline sketches and a measure of size (see below), and including a number of typical profiles. The location of the nucleus, mouth and contractile vacuoles, together with the density, length, width and location of flagella or cilia, should then be added. Separate drawings of details of, for example, the behaviour of the contractile vacuole, the contents of food vacuoles, the patterns of locomotion, the structure of the mouth, and the presence of extrusomes, should also be prepared. Written notes are often very useful.

Drawings need to be accompanied by an estimate of size, which may be made in two ways. The first is to measure the diameter of the field of view (the area that can be seen when looking down the eyepiece) before making observations, by looking at a micrometer slide with a scale (usually 1 mm) etched onto its surface. Measurements of the field of view have to be made for each objective. The size of an organism may then be estimated as a proportion of the field of view. Alternatively, a measuring eyepiece may be used. This contains a scale which is in focus when observing the specimen. Micrometer graticules are inserts which convert normal eyepieces into measuring eyepieces, and they can be bought for most types of eyepieces. Measuring eyepieces have to be calibrated against a micrometer slide. This has to be done for each objective. Organisms are measured as a number of eyepiece units, and this is converted into microns. Multiplying the magnification of the objective and of the eyepiece does not give the magnification of the object being observed.

Although drawings are best made in a firm plain paper notebook, one option is to use large punchcards. The holes may be cut out according to a predetermined code (e.g. to indicate the presence of cilia, flagella or pseudopodia, or to indicate colour, habitats, etc.), and, by using a knitting needle, all previously made drawings with a particular feature can be selected and compared.

Uninterpreted records

The extent to which errors of interpretation of the protozoan form may occur is quite remarkable. Thus, uninterpreted records are highly desirable and ought to be included in professional surveys. Such records can be made by photography, ciné or video. However, since ciné has been rendered obsolete by advancing video technology, it will not be discussed any further. If photography or video are to be used, it is best to have a microscope equipped with separate ports to which cameras may be attached: the usual arrangement is to have a microscope with a trinocular head. Usually, the camera is attached to the vertical tube (Fig. A).

Photomicrography of protozoa requires a camera from which the lens can be removed, an adaptor that will allow attachment to a microscope, and a projection eyepiece. Only if the camera is attached to a separate port will it be possible to make unhindered observations of protozoa while photographs are being taken.

Cameras with a heavy shutter movement (focal plane shutters) will cause vibration in the microscope and movement of the fluid on the slide. Such movement is greatly exaggerated on the film plane because of the magnification factor, and the object will appear blurred or out of focus. It may be necessary to support such cameras on a stand, rather than attach them directly to the microscope. Cameras with diaphragm shutters are to be preferred.

Whereas photography of fixed and stained preparations is straightforward, and any camera with automatic exposure control can be used, fast moving living organisms can only be photographed satisfactorily with an electronic flashgun. There are many cameras on the market that can adjust exposure of subjects illuminated by electronic flashes, but as yet none is available specifically for microscopes. For use in photomicrography, domestic cameras need to be modified: the flash tube must be placed in the light path either by dismantling it from the electronic flash gun and fixing it in an appropriate location in the light path, or by redirecting the flash into the light path of the microscope via a mirror system (Fig. A, Patterson, 1982). Variations and improvements of such systems may be found in the pages of *Microscopy* (the journal of the Quekett Microscopical Club) or in *Mikrokosmos* (in German), both of which cater for the amateur microscopist.

Without an automatic flash exposure system, the exposure will differ with each magnification and with each type of contrast enhancement system.

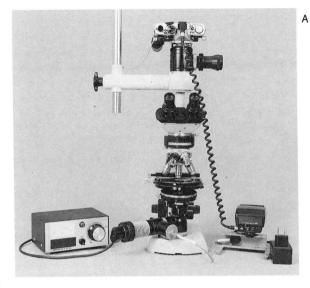

A

Obtaining the correct exposure involves a series of trials with neutral density (grey) filters, such that the light is attenuated to the appropriate intensity.

Photomicrographers should have basic photographic skills and access to a darkroom. Microscopically viewed objects have much poorer contrast than conventional subjects, and specialist films and developing or printing techniques may be needed. Photography of living protozoa has a high film-wastage rate, as the organisms move at inappropriate moments or with unkind rapidity. The best black and white films, for example, *Kodak Technical Pan*, are those that have very high contrast, adjustable sensitivity, and an insignificant grain. For colour work, the author has used conventional colour-reversal (slide) film. Special photomicrographic films are available, but they are difficult to obtain, expensive, and offer marginal improvement over normal films. If an electronic flash is to be used, daylight colour films are appropriate; without a flash, a blue filter or an artificial light film will be needed.

Photographs are often marred by dust within the optical system. The most common sites from which images of dust are projected onto the picture are the glass surfaces between the objective and camera, and the front face of the lamp housing of the microscope. Dust should be removed with compressed air (see p.11). Slight unfocusing of the condenser usually eliminates any images of dust inside the lamp housing.

Video microscopy

Video microscopy is an ideal medium for teaching microscopy and protozoology, as many smaller organisms can be identified from their movements and the images recorded on video tape. VHS tapes give poor-quality results if edited, but S-VHS, BETACAM and U-MATIC formats are suitable if editing is required. The sound-track can be useful for recording verbal comments.

It is now as cheap to buy a colour camera as a black and white one, and colour is recommended. Camcorders can be set to focus on infinity and simply directed down the eyepiece, but the result is far from satisfactory. It is preferable to use surveillance-style cameras without a lens, which are attached to the microscope using a special adaptor. Adaptors may be bought from microscope suppliers.

Where to find and how to collect protozoa

Protozoa occur in trophic or encysted states in virtually any kind of natural habitat which is temporarily or permanently wet. The numbers of active, trophic individuals will be determined by the amount of food available and by the prevalence of predators. Many protozoa directly or indirectly rely on decaying organic (vegetable) debris or on unicellular algae for food. The richest sources of protozoa in natural habitats are sites of high productivity, such as shallow ponds, or the borders of larger standing bodies of water where leaves and other plant matter accumulate and where the sun penetrates in sufficient strength to support algal growth. Any small body of water that has developed a strong colour or a green mat of material (either across the bottom or within the water) will probably prove to have an algal bloom. Protozoa are often present as consumers of these algae.

Very small ponds and puddles are rarely particularly good sources because there may be a paucity of nutrients, or because continual drying and rewetting creates very demanding conditions in which only a few species can survive. Similarly, flowing waters do not usually contain many protozoa. However, protozoa may be found in and around associated plant material.

Protozoan communities are very changeable and when collecting, transporting and maintaining samples, care must be taken to prevent severe changes in the communities. If samples are kept in a cool place in the laboratory, a succession of different organisms will appear over a period of several weeks. Collections should be made in relatively large glass jars and samples should contain some detritus, but soil and mud are best avoided. Sufficient organic matter to form a loose layer at the bottom of the jar should be included.

Most pond/ditch/lake samples will be collected from clean, organically enriched, or anoxic sites; each type of sample has to be handled differently. If the water from the collection site is clean and aerated, a large amount of air should be enclosed with the sample to prevent it becoming anoxic. Samples from anoxic sites (they usually look black or smell strongly) should not be mixed with clean-water samples, as the sulphides that are present will kill the protozoa from aerated sites, and the oxygen will kill protozoa living in anoxic sites. For this reason, one should also avoid mixing air with the sample. Protozoa from organically enriched sites (e.g. polluted sites and sewage works) usually need some oxygen, but as this can be very rapidly depleted by bacterial degradation of organic matter, samples should be stored as a thin layer of fluid with plenty of air.

Open bodies of water are worth sampling, especially if they have a distinctive greenish, olive or brown colour, as this may suggest a bloom of algal growth. However, samples from open waters have to be concentrated; this is usually done in the laboratory (see below), but it can be done on site. The most simple way is to pour the sample through a coffee filter, shake it (before all the fluid has gone) to resuspend collected material, and tip the concentrated sample into a container. Many species die soon after being concentrated. Samples may also be taken with a plankton net, ideally with a mesh size of less than 20 µm if the net is to retain protozoa. Such nets are expensive. A slower but convenient method is to reverse filter a sample. A plastic or glass cylinder with a membrane (e.g. 0.45 µm Nuclepore) filter or a 20 µm (or finer) mesh filter (obtainable from Staniar, see p.220) over one end is placed into a sample. Water passes into the cylinder, but the protozoa are held back by the mesh. The water can be removed from inside the cylinder with a pipette.

Alternatively, sponges or slides inserted into expanded polystyrene foam may be suspended in a water column for a week or so, and communities will develop on these.

If a sample is left to stand undisturbed for several

hours, various species may move towards the light, or accumulate at the top, bottom or sides of the container. Such aggregations are sometimes visible with the naked eye, but flat glass containers (such as Petri dishes) and a dissecting microscope do make it much easier to find them. Generally, samples taken from the bottom of a jar will contain most protozoa.

Muds, sands, peat slurries and other sediments are often rich in protozoa. Samples of the top few millimetres (i.e. without the deeper anoxic material) may be taken with a spoon (wok spoons are ideal). In the laboratory the sediment may be used to inoculate a fluid culture (see below). Most protozoa in sediments are motile and will move upwards; they may be collected by placing the sediment in a dish, removing the excess fluid after a few hours, placing a layer of lens tissue over the mud, and adding some coverslips. Two or three days later, a community of organisms will have developed on the undersurface of the coverslips. Older preparations usually become anoxic.

Soils contain many protozoa, and may be used to inoculate fluid cultures (see below). Virtually every species found in soils has the capacity to encyst, and good results are obtained if the soil is first allowed to dry out completely. Dry crumbs of soil are then used to inoculate fluid or agar cultures. Alternatively, the soil may be dampened so that fluid and associated protozoa may be squeezed out by finger pressure.

Mosses, such as *Sphagnum*, also contain many protozoa, especially testate amoebae and various species with symbiotic algae. Handfuls of the filaments that extend to the ground should be collected, placed in a plastic bag and returned to the laboratory. Protozoa will be found in the water that is obtained by squeezing the sample several times.

Other sites that are rich in protozoa include slimes on solid surfaces over which water is running, areas under ice, and the gutters of houses. Protozoa also occur in abundance in sewage works, where they play an important role in clarifying the water (they remove suspended bacteria). The protozoa are found in the sludge of activated sludge plants, in the organic layer in trickle filters, or on biodiscs.

The relative abundance of protozoa varies according to the time of year. The greatest diversity occurs in late winter and very early spring, when there are relatively few metazoa. Freshwater protozoa with symbiotic algae appear to be relatively abundant early in the year in temperate climates.

Samples that contain large numbers of animals (small crustacea, worms, midge larvae, etc.) will not provide as rich a variety of protozoa as those without metazoa. If metazoa are present, the protozoa will soon disappear after the material has been returned to the laboratory, and it is advisable to pass the sample through a sieve or other crude filter to remove larger organisms if a long-lived sample is required.

Keeping samples

Once collected, protozoan samples should be protected from temperature changes and kept out of direct sunlight. If it is important to report accurately the most common species that are present, a list must be made within 12 hours of collection.

If stored out of direct sunlight (e.g. near a north-facing window in the northern hemisphere), and if protected from cold and heat, samples will usually provide a changing community of protozoa for several weeks. If placed in large, flat containers, the samples can be monitored with a dissecting microscope to establish the diversity and abundance of organisms.

It should be noted that an excess of organic matter can cause cultures to 'go off': the organic matter creates a biological demand for oxygen that cannot be met by diffusion from the surface. The cultures first become milky with bacterial growth, and anoxic (even reduced) conditions may follow. This can be avoided by keeping the amount of organic matter to a minimum and by ensuring that the sample has a large surface area.

Cultures

In order to maintain long-term cultures, it is necessary to provide a medium that suits each species, and a supply of appropriate food. Glassware that has been cleaned in hot water and rinsed of all detergent provides the best culture vessels.

Cultures can be selective or non-selective. The latter are usually samples of water enriched with some food material. Since most protozoa eat bacteria or bacterial detritus, the simplest way to enrich a culture is to add several boiled barley, wheat or rice grains. However, as cultures enriched in this way tend to distort the community structure, this approach cannot be used to provide a list of all the protozoan species initially present in a sample.

Non-selective cultures often produce similar species (equivalent to garden weeds). Genera that commonly emerge in organically enriched cultures include the flagellates *Chilomonas*, *Bodo* and *Para-*

physomonas; and the ciliates *Paramecium, Cinetochilum, Cyclidium, Halteria,* and *Colpidium* (pp. 186–187).

More selective cultures are obtained either by offering food that will suit particular protozoa only, or by collecting one or more individuals of one species and inoculating them into a suitable medium with food. The best medium is filtered fluid from the sampling site, but most freshwater and soil species will also grow in commercially available, non-carbonated spring waters with a low mineral content.

For the most selective cultures, it is necessary to catch individual protozoa with a fine pipette. This can be frustrating at first. It helps to hold the pipette so that it does not shake: it can be braced across several fingers, or held with two hands. The pipette is kept relatively still under the microscope, and the sample is gently moved around to bring the organisms to the pipette, rather than the other way around. Alternatively, a small drop of fluid can be drawn up via a teat or a tube to the mouth. For very small organisms, it may be necessary to carry out this process under a compound microscope, using a mechanically driven (hypodermic) syringe to draw up small quantities of fluid.

If absolute purity is required of a sample, around 20 selected cells are placed in several millilitres of medium. The cells are collected a second time and the process repeated. This 'washes' the protozoa, and also removes any contaminating species. The target species can then be inoculated into fresh medium with a source of food. The most rewarding growth is usually achieved if the initial cultures are of small volume (no more than several millilitres), and it is advisable to build up the size of the cultures gradually.

The technique of inoculating cultures with small numbers of selected cells often fails because the right kind of food is not available, or because the composition of the medium is not ideally suited to the target organism. Some species grow best in mixed cultures, and this is especially true of larger genera such as *Amoeba* or *Stentor*.

Some protozoa do not grow well, or conveniently, in a fluid medium, being more suited to thin films of water. They can be grown on solidified 1.5–2 % agar: 1.5 g of agar are added to 100 mls of medium, placed in a boiling water bath until molten, poured into shallow (usually Petri) dishes, and left to gel. Agar is suitable for many amoebae and soil protozoa, and some flagellates. The dishes should be covered to prevent evaporation while the protozoa grow. If used with soils, this approach is particularly successful in cultivating small amoebae. Fluid cultures of soils tend to produce the ciliate *Colpoda*.

Most protozoa are selective feeders and cultures must seek to provide appropriate food. The principal categories of food comprise bacteria in suspension, bacteria adhering to surfaces, other protozoa, algae, and detritus. The simplest organisms to culture are often those that eat bacteria, a supply of which can be guaranteed by adding boiled barley, wheat, or rice grains to support bacterial growth.

Other media and methods of culture are to be found in, for example, Finlay *et al.* (1988), Kirby (1941) and Lee *et al.* (1985).

Classification of protozoa

Classification schemes for organisms fulfil two functions:
- A filing system from which data may be conveniently retrieved.
- A means for expressing ideas about evolution.

However, because ideas about patterns of evolution are always changing, classification schemes are inevitably unstable. This is especially true at the moment for protozoa. Consequently, given below is a short list of the major types of protozoa (along with a few distinguishing features), which is intended to be a simple filing system; evolutionary relationships are not implied. It should also be noted that some groups appear more than once, and that included in each group are those species that are illustrated or mentioned in this book.

1 CILIATES: protists with cilia in lines (kineties) at some stage in the life cycle. Two kinds of nuclei.

Chonotrichs: ectosymbiotic ciliates with a spiral fold of cytoplasm around the unattached end. *Spirochona.*

Colpodids: mostly filter-feeding ciliates, using tightly packed feeding cilia clustered around the mouth. No undulating membrane. Somatic cilia arranged in pairs. *Bursaria, Colpoda, Cyrtolophosis.*

Cyrtophores: motile ciliates with a cluster of strongly developed microtubular rods or nematodesmata, normally used for manipulating algae or large lumps of debris into the mouth. Either flattened (hypostomes) or rounded. *Chilodonella, Chlamydodon, Drepanomonas, Nassula, Phascolodon, Pseudomicrothorax, Trithigmostoma, Trochilia.*

Haptorids: predatory ciliates with an armoury of killing and/or holding extrusomes around the mouth region. The mouth may be apical or arranged along one flattened margin of the cell. *Amphileptus, Chaenea, Didinium, Dileptus, Homalozoon, Lacrymaria, Litonotus, Loxophyllum, Monodinium, Phialina, Spathidium, Trachelius, Trachelophyllum.*

Karyorelicts: ciliates with non-dividing macronuclei. *Loxodes.*

Oligohymenophora: ciliates with a specialized buccal ciliature comprising only three membranelles (blocks of cilia) and an undulating membrane. These organelles can nevertheless be difficult to see. Mostly filter feeders, eating bacteria. Common.
(i) Hymenostomes: oligohymenophora with short membranelles and an undulating membrane. Mouth usually small and difficult to see. Common, especially in organically enriched sites. *Cinetochilum, Colpidium, Glaucoma, Tetrahymena.*
(ii) Scuticociliates: oligohymenophora in which the undulating membrane is typically a long and well-developed, veil-like structure to the right of the mouth. *Calyptotricha, Cohnilembus, Cyclidium, Lembadion, Pleuronema.*
(iii) Peritrichs: oligohymenophora with buccal ciliature forming one or more wreaths around the broad anterior part of the cell. Usually bell-shaped. Mostly sessile. *Astylozoon, Carchesium, Cothurnia, Epistylis, Hastatella, Orbopercularia, Opercularia, Ophrydium, Platycola, Rhabdostyla, Vaginicola, Vorticella.*
(iv) Peniculines: oligohymenophora in which the membranelles are drawn out as relatively elongate structures. Undulating membrane weakly developed. Usually with trichocysts and star-shaped contractile vacuole complexes. *Frontonia, Paramecium, Neobursaridium, Urocentrum.*

Polyhymenophora (spirotrichs): ciliates that feed using a band of membranelles stretching from the anterior pole of the cell to the cytostome. The band is called the adoral zone of membranelles (AZM).
(i) Hypotrichs: polyhymenophora that walk on the substrate using cirri (blocks of cilia). Usually dorsoventrally flattened. *Amphisiella, Aspidisca, Chaetospira, Euplotes, Holosticha, Oxytricha, Paruroleptus, Pattersoniella, Stichotricha, Stylonychia, Tachysoma, Uroleptus, Urostyla.*
(ii) Heterotrichs: polyhymenophora that move with somatic cilia arranged in kineties. *Blepharisma, Brachonella, Caenomorpha, Climacostomum, Condylostoma, Metopus, Spirostomum, Stentor.*
(iii) Oligotrichs: polyhymenophora in which the somatic cilia are absent or reduced to a circumferential band of spines. AZM is apical and well developed. Mostly open-water organisms. *Halteria, Strombidium, Strobilidium, Tintinnidium.*
(iv) Epalcids: polyhymenophora with a flattened sculpted body. Somatic cilia reduced or absent. AZM usually near the middle of the body and reduced. Mostly from anoxic sites. *Discomorphella, Epalxella.*

Prostomes: ciliates with an apical mouth (normally quite distensible) used mostly for ingestion of debris, detritus, damaged cells or tissue. Mostly associated with detritus. *Coleps, Mesodinium, Urotricha, Prorodon.*

Suctoria: ciliates without cilia during the trophic stage. This stage is usually sessile and immotile, and feeding is by means of one or many radiating arms (= mouths), each of which is equipped with holding extrusomes at its tip. *Acineta, Dendrocometes, Podophrya, Tokophrya, Trichophrya.*

2 FLAGELLATES: protists with 1–8 flagella, usually located apically or subapically. With chloroplasts or without.

Bicosoecids: sessile cells of a single genus. Two flagella insert anteriorly, but one is directed backwards (recurrent) and attaches the cell to the bottom of a vase-shaped, organic lorica. Eat suspended bacteria. Without chloroplasts. *Bicosoeca.*

Bodonids: small, biflagellated organisms. Flagella insert subapically or laterally, with one directed laterally or anteriorly, and one recurrent. One genus is attached; the others usually move by gliding or skipping. Usually eat individual adhering bacteria, taken in via a discrete mouth. Without chloroplasts. *Bodo, Cephalothamnium, Rhynchomonas.*

Cercomonads: gliding flagellates with two flagella, one of which trails on the ground, often with cytoplasm being pulled out behind the cell, and one active anterior one. Colourless. Feed on bacteria by pseudopodial engulfment. *Cercomonas, Heteromita.*

Chrysophytes = Chrysomonads: cells with two flagella: typically, one is short and flaccid, and the other is longer. Either with chloroplasts (golden) or without, capable of phagocytosis or not, with or without a layer of surrounding siliceous plates, sessile or motile, and solitary or colonial. *Anthophysa, Chrysamoeba, Chrysosphaerella, Dendromonas, Dinobryon, Mallomonas, Ochromonas, Paraphysomonas, Poterioochromonas, Spumella, Syncrypta, Synura, Uroglena.*

Collar flagellates: with single apical flagellum, around which lies a collar comprised of fine cytoplasmic 'fingers'. Mostly sessile. May be solitary or colonial, naked or loricated. Without chloroplasts. Eat by filtering suspended bacteria or other small particles. *Codonosiga, Diploeca, Diplosigopsis, Monosiga, Pachysoeca, Sphaeroeca.*

Cryptophytes = Cryptomonads: rigid cells with two flagella arising within an anterior depression. The depression is lined with ejectisomes. Either with chloroplasts (off-green, orange, blue-green, red) or colourless (in which case osmotrophic). *Chilomonas, Cryptomonas, Cyanomonas, Goniomonas.*

Dinoflagellates: rigid cells with two flagella: one passes horizontally around the body, usually in a groove; the other passes longitudinally, often trailing behind the cell. Rounded. May occasionally be drawn out into spines. With (orange or off-green) chloroplasts or without. Colourless species are osmotrophic, consume detritus, or prey on other protists. *Gymnodinium, Gyrodinium, Ceratium, Amphidinium, Peridinium.*

Dilpomonads: cells with two nuclei and two clusters of four flagella, each arising at the head of a longitudinal groove, with some flagella projecting laterally and others trailing behind. Usually from anoxic or organically enriched sites. Without chloroplasts. Osmotrophic or eating bacteria. *Trepomonas, Hexamita.*

Euglenids: small to medium-sized cells with (usually) two flagella arising in an anterior flagellar pocket, both, one or none of which may emerge. Move by swimming or gliding, or squirming. Body may be pliable. Some with ingestion apparatus. Eat bacteria, detritus or other protists. Some have one or more bright green (chlorophyll b present) chloroplasts. *Anisonema, Astasia, Entosiphon, Euglena, Heteronema, Menoidium, Notosolenus, Peranema, Petalomonas, Phacus, Trachelomonas, Urceolus.*

Heterolobosea: protists with an amoeboid and a flagellated stage in the life cycle. The flagellated stage has two or four flagella, a flexible body, and usually does not eat. Includes facultative pathogens. *Naegleria.*

Others: there are about 70 genera of heterotrophic flagellates that cannot be confidently assigned to any of the groups of flagellates listed in this Guide (Patterson and Larsen, 1991). They include *Artodiscus, Clautriavia, Helkesimastix, Kathablepharis,* and *Multicilia.* In addition, oomycete fungi, protostelid and eumycetozoan (e.g. *Ceratiomyxa*) slime moulds, and desmothoracid heliozoa (among others) produce flagellated organisms as part of their life cycle.

Pedinellids: cells with a single apical flagellum. Usually with a stalk, although they may swim trailing the stalk behind. When sessile, the flagellum is surrounded by a small number of discrete arms which may be used to intercept particles of food. *Actinomonas, Pteridomonas, Ciliophrys.*

Pelobionts: amoeboid cells, usually with a single, relatively long, stiff flagellum. From anoxic sites. *Mastigamoeba, Mastigella, Pelomyxa.*

Phalansteriids: flagellates with a single flagellum

that has a tight basal collar. Live in colonies supported by globular mucus. Eat bacteria. *Phalansterium*.

Pseudodendromonads: Biflagellated cells. The two similar flagella are located at one edge of flattened triangular cells. One genus colonial, supported on a stiff, dichotomously branching stalk system. Eat bacteria and other small suspended particles. *Pseudodendromonas*.

Spongomonads: biflagellated cells living in colonies supported by globules of mucus. *Rhipidodendron, Spongomonas*.

Volvocales: flagellates with two or four apical flagella. Most species have a bright green chloroplast containing chlorophyll b. With rigid cellulosic wall. Often forming rounded, swimming colonies. *Brachiomonas, Carteria, Chlamydomonas, Chlorogonium, Eudorina, Gonium, Haematococcus, Pandorina, Polytoma, Polytomella, Volvox*.

3 AMOEBAE: traditionally, organisms that move and/or feed using temporary extensions of the cell body (pseudopodia). Usually include star-like protists with stiffened pseudopodia (the heliozoa), and rhizopod amoebae, which produce short-lived pseudopodia.

Actinophryid: heliozoan body form. Arms taper from base to tip. Two genera only: one has a single central nucleus; the other has a layer of nuclei underlying a layer of vacuoles. *Actinophrys, Actinosphaerium*.

Centrohelid: heliozoan body form. Arms are relatively thin and do not taper. Extrusomes prominent. Microtubular supports terminate on a central granule. With or without a layer of scales or spines on the body surface. *Acanthocystis, Chlamydaster, Heterophrys, Oxnerella, Raphidocystis, Raphidiophrys*.

Desmothoracids: sessile, heliozoan-like cells living within a perforated lorica, out of which the arms project. With a conventional amoeboid and flagellated stage in the life cycle. *Clathrulina, Hedriocystis*.

Diplophryids: cell body enclosed in a delicate organic shell. With two tufts of fine pseudopodia emerging from opposing ends of the cell, and with one or more large orange lipid droplets. From organically enriched sites. *Diplophrys*.

Euamoebae: rhizopod amoebae with one or more broad pseudopodia, and without a firm shell. May have short, stubby, filose subpseudopodia emerging from a larger pseudopodial region. Pseudopodia usually develop either gradually (progressive) or suddenly (eruptive). Some species have a flagellated stage; others are facultative pathogens. *Acanthamoeba, Amoeba, Astramoeba, Cashia, Chaos, Cochliopodium, Hartmanella, Mayorella, Saccamoeba, Thecamoeba, Vannella*.

Heterolobosea: protists with amoeboid and flagellated forms. Amoebae usually small (less than 50 µm), with eruptive bulging of the pseudopodia. Includes facultative pathogens. *Naegleria*.

Leptomyxids: naked amoeboid organism, with cytoplasm forming anastomosing channels. *Leptomyxa*.

Nucleariid filose amoebae: with thin pseudopodia, usually arising at any part of the body. Flattened or spherical. Naked or with mucus sheath, or with adhering siliceous particles. No extrusomes (or extrusome-like granules) on the pseudopodia. *Nuclearia, Pompholyxophrys, Pinaciophora*.

Others: there are numerous amoeboid organisms that have yet to be properly described. Those least studied have very thin pseudopodia, either bearing extrusomes (*Gymnophrys, Biomyxa, Microcometes, Reticulomyxa*) or smooth (*Belonocystis, Elaeorhanis*). Some large, shelled species (*Allelogromia, Gromia, Lieberkuhnia*) may be related to marine foraminifera and are in need of further study.

Pelobionts: typically amoeboid organisms with a single long flagellum. However, in one genus (*Pelomyxa*) the flagella are very difficult to see and so this genus has usually been described as an amoeba. *Mastigamoeba, Mastigella, Pelomyxa*.

Testate amoebae: amoeboid organisms with a shell of organic matter, or adhering particles around the body. Pseudopodia emerge from one or two apertures. Either with filose pseudopodia (*Amphitrema, Assulina, Cyphoderia, Euglypha, Trinema*) or with lobose pseudopodia (*Arcella, Cyphoderia, Centropyxis, Cryptodifflugia, Difflugia, Lecquereusia, Nebela, Quadrulella*).

Vampyrellids: flattened amoeboid organisms with cell margins giving rise to numerous very delicate pseudopodia. Often eat algae or fungi, and often orange in colour, with a granular texture to the cytoplasm. *Arachnula, Vampyrella*

4 NON-PROTOZOAN TAXA INCLUDED:

Algae: protists with chloroplasts. Only those without flagella are included here; the remainder are listed under 'Flagellates' (above). Mostly unicellular organisms are included here, but many algae are multicellular.

(i) Diatoms: with a siliceous wall and orange or golden chloroplasts. Filamentous or solitary, motile or immotile, centric (pill-box shape) or pennate (cell with discrete poles). *Melosira, Navicula, Nitzschia, Pinnularia, Stephanodiscus, Tabellaria.*

(ii) Green algae (Chlorophyta): cells with cellulosic cells walls and bright green (chlorophyll b) chloroplasts. With a rich variety of shapes (e.g. coccoid and filamentous). *Closterium, Eurastrum, Hyalotheca, Micractinium, Micrasterias, Mougeotia, Spirogyra, Spondylosium.*

Prokaryotes: (bacteria). Without internal organelles, but may have inclusions. Typically with one dimension restricted to about 1 μm. Usually distinguished by colour (e.g. blue-green algae), form or metabolic pathways (e.g. spiral, coccoid, filamentous, red sulphur, etc.).

Blue-green algae: (cyanobacteria). Prokaryotic organisms carrying out photosynthesis. Either solitary cells or filamentous – the latter are more obvious. A type of bacterium, but forms an ecologically distinct group. Some are endosymbionts in protists. *Oscillatoria.*

Metazoa: organisms with many cells which are arranged in epithelia (layers attached to collaginous sheets). Different cells may have different functions.

(i) Gastrotrichs: pliable bodies, usually with locomotor cilia, a mouth that opens anteriorly, and two posterior adhesive structures. *Chaetonotus.*

(ii) Rotifers: cilia usually reduced to two anterior clusters used for feeding. Body surface rigid and normally made of articulating elements. Usually has a mastax located near the anterior end of the digestive system, and two posterior adhesive structures. *Polyarthra, Squatinella.*

(iii) Nematodes: elongate stiff bodies that are rounded in cross section. Move by writhing or serpentine gliding through the substrate. Typically with a muscular pharynx near the front end.

(iv) Flatworms: very pliable bodies. Move using a combination of muscular activity and the cilia which cover the entire body. Typically with a muscular pharynx opening on the ventral surface, away from the anterior end of the cell.

(v) Tardigrades: rigid bodies with an exoskeleton made of a number of articulating elements. With eight stubby legs ending in claws. *Macrobiotus.*

How to use the key

What follows is a simple, dichotomous key. At any given step (e.g. Step 1) you are presented with a pair of options (A and B). Decide which statement best fits the organism you are looking at. At the end of the option, you are advised to go to another step. There you repeat the process, continuing until you are provided with a generic name instead of being directed to another step. An illustration is provided so that you can check whether you have the correct organism.

Unfamiliar objects can only be understood with the acquisition of new concepts and new terms. Thus, successful use of the key will involve a learning process and necessitate effort from the user. The key is followed by a glossary to ease the process.

The key makes little provision for cells which are damaged, for example, through being squashed. You should therefore try to find several individuals of each species, and it should then be possible to distinguish normal features (present in all cells) from those which are abnormal (peculiarities of individual cells). A good example of the need for careful observation occurs where the key asks if the cells swim or glide. If one cell only was observed, and this has been compressed, such a question cannot be answered reliably.

This key does not claim to be comprehensive. You may encounter organisms that are not included. The drawings, which have been simplified for clarity, may generalize the characteristics of a genus. However, the true appearance of organisms will be evident from the photographs.

The Key

The purpose of this first step is to draw attention to those protozoa that do not have the obvious characteristics of flagellates, ciliates, or amoebae, and to certain other organisms that may be confused with protozoa.

A Unicellular organisms with nuclei, that move or feed using flagella (flagellates), cilia (ciliates) or pseudopodia (amoebae). They may occur singly or in colonies; they may swim freely, glide in contact with a substrate, or be sedentary; they may be housed in a lorica (test or shell), clothed in scales or other adhering matter, or be naked; they may or may not be coloured. Most are 5–1000 μm in length. PROTOZOA. Some basic types are illustrated in Figs 1–3.

GO TO 2

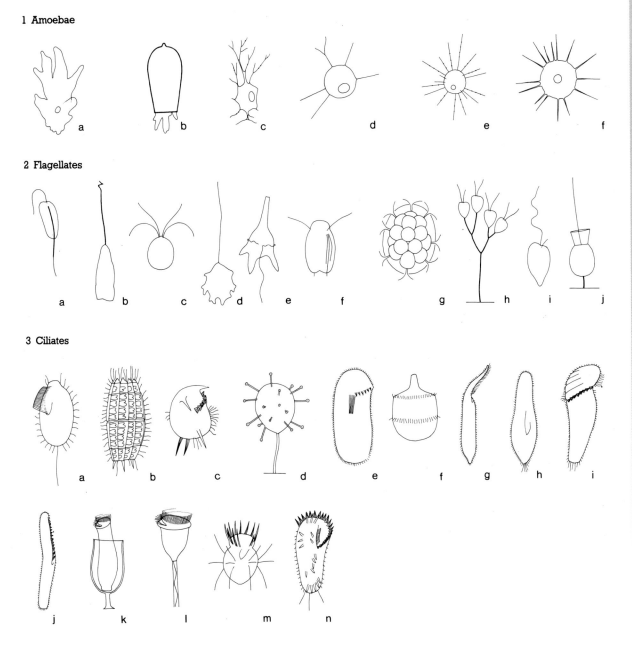

1 Amoebae

2 Flagellates

3 Ciliates

Protozoa are divided into three major morphological types: flagellates (Step 2), amoebae (Step 72) and ciliates (Step 116). All protozoa that move actively, or that create currents of water while feeding, do so with cilia or flagella. These are virtually the only actively moving organelles to be found in eukaryotic cells. Cilia and flagella are difficult to see, especially if only bright-field optics are available (see methods, p. 13). Flagella are as long as or longer than the length of the cell and are few (eight or fewer) in number. Cilia are more numerous and relatively short.

Amoebae move or feed by means of temporary extensions of the cell surface (pseudopodia) (Figs 1(a)–(f) & 139). Pseudopodia may also be produced by some flagellates, usually when feeding. Certain ciliates may appear amoeboid as they squeeze through small spaces, so care is needed to distinguish flexibility from the capacity to produce pseudopodia. Protozoa squashed under a coverslip may become distorted; again care must be taken not to mistake the distortion for pseudopodia.

Protozoa that are not very motile may easily be overlooked. Some amoebae, especially those that are shelled (testate or loricated), withdraw their pseudopodia. Naked amoebae may adopt a radiate floating form (Fig. 142) if disturbed. Heliozoa (Figs 1(e) & (f), Step 187) have stiff radiating arms and often appear inactive, although, in fact, they move very slowly. Trophic suctoria (Fig. 3(d), Step 195) are attached. They are a kind of ciliate, have stiff arms and do not move at all.

General guides to protozoa include Sleigh (1989), Lee *et al.* (1985), Margulis *et al.* (1990), Kudo (1966), Streble and Krauter (1981), and Grassé (1952, 1953, 1984). For guidance on the ecological literature, see Finlay and Ochsenbein-Gattlen (1982) and Fenchel (1987).

B Small organisms that move around, but cannot be confidently assigned to any of the types in Figs 1–3.

A variety of non-protozoan organisms are about the same size as protozoa (5–1000 µm) and can easily be confused with them. Except for unicellular algae that are related to protozoa, most are not keyed out here. Organisms that may cause confusion include prokaryotes (bacteria and blue-green algae) and eukaryotes. Eukaryotes can be distinguished by discrete compartments (e.g. nuclei, chloroplasts, vacuoles) inside the cytoplasm. If coloured and gliding, the organisms are most likely to be prokaryotic blue-green algae (Fig. 4), or pennate or filamentous diatoms (Figs 5, 6, 7). Barely motile and non-motile algae include centric diatoms (Fig. 8), desmids (Figs 9, 10 & 11), and some coccoid eukaryotic green algae (Fig. 12). Non-green gliding organisms include sulphur bacteria, often with pink, refractile inclusions (Fig. 13). Smaller metazoa (animals, Fig. 14) with cilia, such as rotifers (Fig. 15), gastrotrichs (Fig. 16) and flatworms (Fig. 17) may be confused with ciliates. Most can be distinguished from protozoa because they have a discrete gut or muscular pharynx. Other small metazoa include the nematodes (Fig. 18) and tardigrades (Fig. 19). Further information on these groups may be found in the following general accounts: Streble and Krauter (1981), Ward and Whipple (Edmonson, 1969), and Pennak (1989), for bacteria, protists and small invertebrates; bacteria are reviewed by Starr *et al.* (1981) and Holt (1984–1989); blue-green algae are discussed in Bourrelly (1970), and diatoms in Barber and Haworth (1981) and Round *et al.* (1990). Sims (1980a & b, 1988) and Kerrich *et al.* (1978) review the available invertebrate identification guides. Specific accounts of rotifers are found in Pontin (1978), and flatworms in Young (1970, 1972).

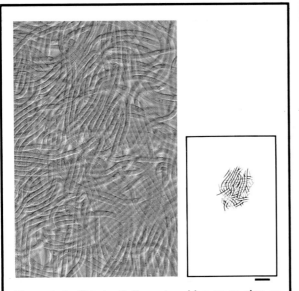

Figure 4 *Oscillatoria*. A filamentous blue-green alga or cyanobacterium. Photosynthetic pigments include phycobilins, which give these prokaryotic organisms their characteristic bluish tinge. As bacteria, these algae have no internal organelles. The filaments are comprised of many disc-shaped cells joined end to end. The cells can glide. Some filamentous blue-green algae have occasional swollen cells, called heterocysts. *Differential interference contrast.*

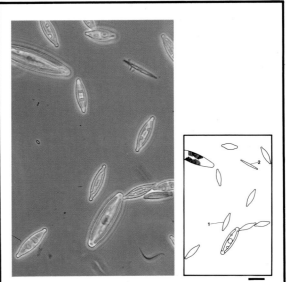
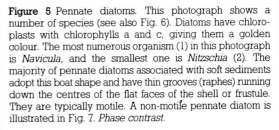

Figure 5 Pennate diatoms. This photograph shows a number of species (see also Fig. 6). Diatoms have chloroplasts with chlorophylls a and c, giving them a golden colour. The most numerous organism (1) in this photograph is *Navicula*, and the smallest one is *Nitzschia* (2). The majority of pennate diatoms associated with soft sediments adopt this boat shape and have thin grooves (raphes) running down the centres of the flat faces of the shell or frustule. They are typically motile. A non-motile pennate diatom is illustrated in Fig. 7. *Phase contrast.*

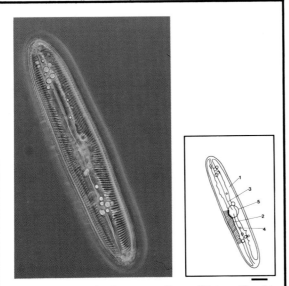

Figure 6 *Pinnularia*. A pennate diatom. Diatoms like this (see also Fig. 5) are common in soft sediments. They have golden chloroplasts (1). The cells are enclosed within a siliceous shell or frustule, which is sculpted with ridges, grooves, and lines or holes. These appear as striae (2). The raphe (3) is involved in motion, as these cells can glide slowly across the substrate. Also evident are oil droplets (4), which are used as an energy reserve, and the central region houses the nucleus (5). *Phase contrast.*

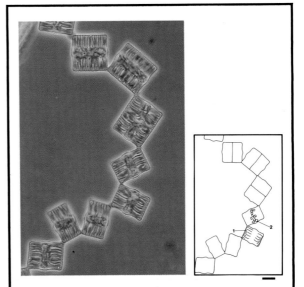

Figure 7 *Tabellaria*. A colonial diatom with a sculpted (siliceous) cell wall (1). The enclosed golden chloroplast (2) belies the presence of chlorophylls a and c. The cells are joined together at their corners and are usually planktonic. *Phase contrast.*

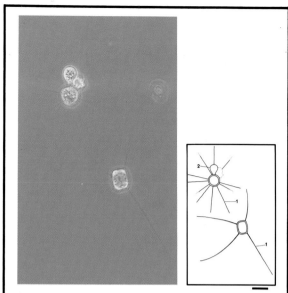

Figure 8 *Stephanodiscus*. A planktonic centric diatom. The cells are solitary, and have a pillbox-like appearance. The lower cell is seen from the side, the upper cell from end on. The margins of the valves give rise to long organic hairs (1). These may reduce sinking rates, thereby minimizing the demands on the energy budget from the flotation system. The upper cell is parasitized by a number of chytrids (2), flagellates that are related to the true fungi. *Phase contrast.*

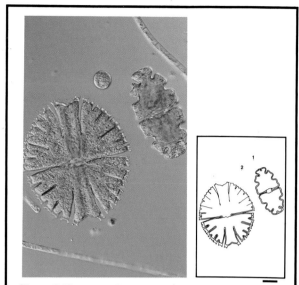

Figure 9 *Eurastrum* (1) and *Micrasterias* (2). Green algae. Like other chlorophytes, they have a rigid external wall made of cellulose, and bright green chloroplasts which contain chlorophyll b. These desmids have no flagella, but are capable of a very slow movement. They are the same size as many motile protists, and are often encountered using the sampling and observation techniques employed for protozoa. Most desmids look like two mirror-image cells joined together. *Differential interference contrast.* (Photo Helge Thomsen.)

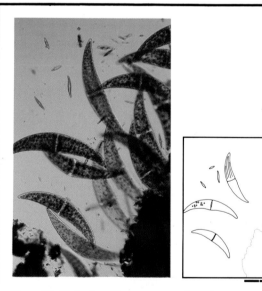

Figure 10 *Closterium*. These crescent-shaped cells are desmids (see also Fig. 9). Desmids are a type of green (chlorophyte) alga, sharing with other green algae the presence of an external cellulosic cell wall and chloroplasts with chlorophyll b. Like most desmids, they look as if two mirror-image cells are joined together. They may move very slowly. Some pennate diatoms can be seen in the background. *Differential interference contrast.* (Photo Helge Thomsen.) (Scale bar 100 μm.)

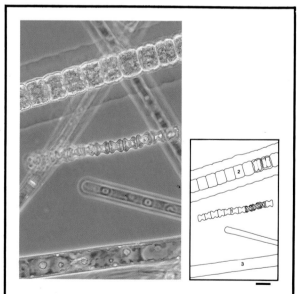

Figure 11 Filamentous algae. This photograph is included to illustrate some of the genera of green algae (chlorophyll b and cellulosic cell walls) that adopt filamentous strategies. The central, short filament of dumbbell-shaped cells (1) is a desmid (see Figs 9 & 10) called *Spondylosium*. Above it is a green filament (2) called *Hyalotheca*, enclosed in a thick transparent sheath. Towards the bottom is a filament of *Mougeotia* (3), the filaments of which pair up for conjugation (as do those of *Spirogyra*) to give a widely spaced crisscross appearance. *Phase contrast.*

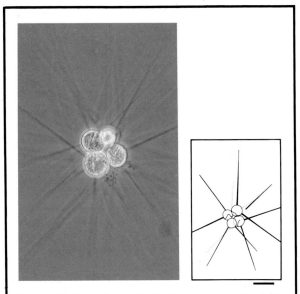

Figure 12 *Micractinium*. A planktonic green alga. Like other green algae, they are bright green because they have chloroplasts that contain chlorophyll b. They have cellulosic cell walls. The cells are not flagellated and are non-motile. The walls are drawn out as long threads, as is often the case with planktonic protists (see Fig. 8). This organism is included because confusion with heliozoa is possible. *Phase contrast.*

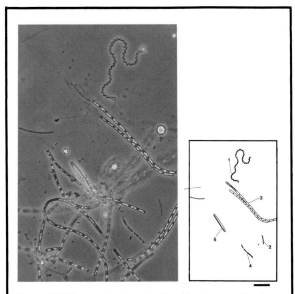

Figure 13 Filamentous bacteria. Included are a motile, flexible, spiral bacterium (1), some colourless rod bacteria (2) and two kinds of filamentous sulphur bacteria (3 & 4). The sulphur bacteria are distinguished by the presence of refractile granules of sulphur that have been deposited within the cells. These organisms live in environments that contain free hydrogen sulphide (smelling like bad eggs), which they oxidize to produce elemental sulphur. The sulphur is deposited within the body. Filamentous sulphur bacteria can usually glide. A pennate diatom (5) is present. *Phase contrast.*

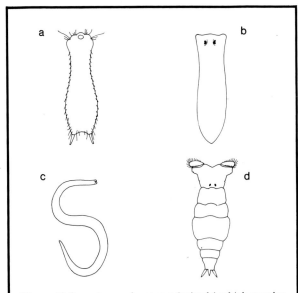

Figure 14 Some types of metazoa (animals) which may be under 1 mm in length and which may be confused with protozoa: (a) gastrotrich; (b) flatworm (platyhelminth); (c) nematode; and (d) rotifer.

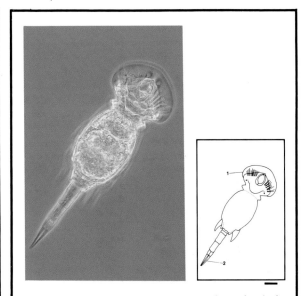

Figure 15 Rotifer. Rotifers are metazoa and are often in the same size range as ciliates, with which they may compete. Rotifers have anterior aggregates of cilia, used in the collection of food; the food is then passed into the gut via two stout grinding plates called the mastax. Most rotifers have a posterior adhesive podite (2). The genus illustrated (*Squatinella*) is planktonic. The anterior end of the cell is atypically developed into a shield-like structure (1). *Phase contrast.*

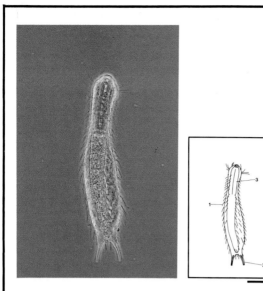

Figure 16 Gastrotrich. Gastrotrichs are a group of poorly understood metazoa. Most freshwater species are very small. Their size range overlaps that of ciliates, with which they can be confused because of the cilia and spines (1). Gastrotrichs can be distinguished by the two adhesive structures at the posterior end of the cell (2) and by the presence of a discrete pharynx (3). They usually glide rather than swim. *Phase contrast.*

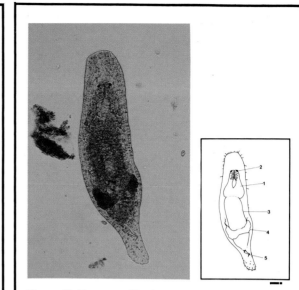

Figure 17 Flatworm (Platyhelminth). The flatworms resemble some ciliates in that they have pliable bodies and a surface coating of cilia. It can sometimes be difficult to distinguish these metazoa from ciliates as they may be in the same size range and because the boundaries between the component cells are usually not easy to see. Flatworms rarely swim, tending to glide against the substrate. The presence of a discrete pharynx (1), eyes (2), gut (3), other internal organs (4, 5), or muscular writhing of the body helps to distinguish these organisms as metazoa. *Bright field.* (Scale bar 100 μm.)

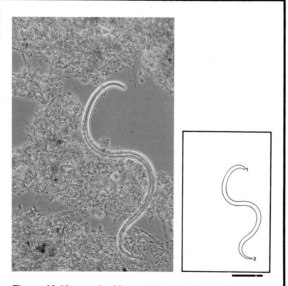

Figure 18 Nematode. Nematodes are extremely common and widespread metazoa. Most nematodes have a long thin shape and a slightly blunt anterior end (1). They are rather stiff, and can move either by writhing (lashing) or by gliding through the substrate. At higher magnifications, a strong muscular pharynx can be seen near the front, and egg-bearing ovaries near the back end (2). *Phase contrast.* (Scale bar 100 μm.)

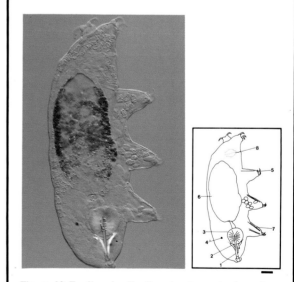

Figure 19 Tardigrade. Tardigrades (or water bears) are characteristic of temporary puddles, mosses, etc. (they have remarkable abilities to withstand desiccation). Tardigrades are metazoa that overlap in size with protozoa, but they are easy to distinguish from the protozoa. Tubular pharynx (1), stylus (2), sucking pharynx (3), eyespot (4), claws (5), intestine (6), claw muscles (7), cloaca (8). *Differential interference contrast.*

A   Cells with flagella.   THE FLAGELLATES   GO TO 3

Flagella are usually parallel-sided structures, about 0.5 μm in diameter, which emerge near or at the anterior of the cell, do not change length and move actively. One or two are usually visible, but sometimes there may be four or more (Fig. 2). Normally at least one is as long as the cell, typical lengths being 5–20 μm). They may beat with a sine wave, in breaststroke fashion, or be held stiffly, sometimes only moving at the tip. Actively beating flagella often move too fast for the eye to see clearly, and only the envelope of the beating pattern can be seen. In gliding flagellates, the flagella may be atypically thick, and adhere to the ground. If there is more than one flagellum, they may be unequal in length, and extend forwards, sideways or backwards, sometimes trailing along the ground.

Some flagellates can produce pseudopodia, usually in order to ingest particles of food. Flagellates with chloroplasts are usually called algae and Bourrelly (1968, 1972, 1985) provides comprehensive guides to the pigmented flagellates and to some groups of colourless ones. All genera of free-living heterotrophic flagellates are reviewed in Patterson and Larsen (1991), and briefer accounts may be found in Hänel (1979), Starmach (1980) or Larsen and Patterson (1990). Prescott (1978) deals more exclusively with algal flagellates.

B   Cells without flagella.   GO TO 72

This step leads to the ciliates (Fig. 3, Step 116) (with cilia, more numerous and shorter than flagella), amoebae (Fig. 1, Step 72), suctoria (Fig. 3(d), Step 196) or heliozoa (Figs 1(e) & (f), Step 188). Some types of flagellate shed their flagella if squashed badly or disturbed. This particularly applies to dinoflagellates (Step 67) and euglenids (Step 61). Some euglenids, typically from sediments, live without flagella (and are not keyed out here).

A   Cells that are attached firmly to the substrate, usually by a stalk, and do not easily detach.   GO TO 4

Most flagellates in this category are colourless, and may either be solitary or live in colonies. Most of the attached flagellates eat bacteria, i.e. they are bacterivorous. Some flagellates do not secrete a stalk, but attach themselves temporarily to the substrate, using the posterior of the cell, e.g. *Ochromonas* (Step 68), *Paraphysomonas* and *Spumella* (Step 50). In these organisms, the body may be drawn out as a fine thread because of the pull from the flagellum. Some species, e.g. *Bodo saltans* (*Pleuromonas jaculans*), attach temporarily by one flagellum. These flagellates key out as motile or attached organisms.

B   Flagellated cells that move freely through the fluid or glide over the substrate. They may sometimes adhere to a surface, but easily detach from it.   GO TO 19

This step leads to algal (with plastids) and protozoan flagellates. However, flagellated swarmers of some fungi, slime moulds (Fig. 20), and amoebae may also be encountered.

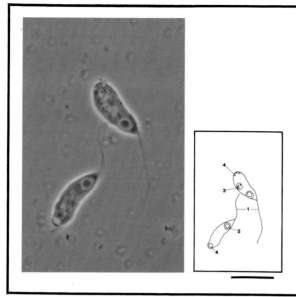

Figure 20 *Ceratiomyxa* swarmers. *Ceratiomyxa* is a slime mould, the trophic stage of which is amoeboid (plasmodial). The organism also produces aggregates of cysts on the tip of a stalk, a trait not dissimilar to that of some fungi; it is from this similarity that the term 'mould' derives. However, the organisms are capable of transforming into flagellated 'swarmers' which have 1–4 flagella (1). Both cells illustrated have a single flagellum; this inserts at the apex of the cell and appears attached to the nucleus (2). Food vacuoles (3) contain bacteria, indicating that the swarmers feed. The contractile vacuole complex (4) usually occurs at the posterior end of the cell. This has been included to show that protozoan-like cells may simply be one stage in the life cycle of another organism. These swarmers are most likely to be confused with some mastigamoebae (see Fig. 86). *Phase contrast.*

4
(3)

A Cells that occur singly or in irregular groups.

GO TO 5

Solitary flagellates, if grouped together, may be distinguished from colonial species because they are not joined by common stalks, common lorica (test or shell) material or common cytoplasm, and do not form regular arrays. Note that new colonies of many colonial species are started with single cells.

B Cells living in colonies.

GO TO 12

5
(4)

A Cells without chloroplasts.

GO TO 7

B Cells with chloroplasts.

GO TO 6

The chloroplasts of most groups of algae have a distinctive colour because of the combination of photosynthetic pigments. Colour can aid identification. Common colours are: bright green (chlorophyll b) (volvocids and euglenids); or golden or off-green (chrysophytes, cryptophytes and dinoflagellates) (see Figs 59, 119, 126, 129 & 132 for a comparison of these colours). Among other coloured inclusions are: eyespots or stigmata (usually red or orange); endosymbiotic algae (green or blue-green), or partly digested residues of food. A common problem with some microscopes is chromatic aberration, as a result of which refractile particles may appear to be green. This is more evident if the condenser iris is closed or the condenser lowered. To establish whether coloration is natural, the microscope should be set up for optimal illumination (see Introduction), and all irises opened. Chromatic aberration will then be minimal and, if present, photosynthetic pigments only will be seen.

A Cells with a ring of fine stiff tentacles around the single anterior flagellum. The flagellum beats with a sine wave in a single plane. It draws a current of water towards the unattached end of the cell. With a thin stalk, and six golden chloroplasts. Mostly 10–20 μm.   Fig. 21 PSEUDOPEDINELLA

There are several genera of pedinellids with chloroplasts (Bourrelly, 1968; Zimmermann *et al.* 1984). Most are encountered swimming in large arcs.

When swimming, or if shocked, the arms may be withdrawn. For colourless relatives, see Step 9. Usually allied with the chrysophytes.

B Cells with one or two chloroplasts and two flagella. One flagellum is long and held in a slight arc; the second may be difficult to see as it is short and curves back over the cell. Mostly under 10 μm.
Fig. 22 POTERIOOCHROMONAS
OCHROSTYLON (not illustrated)

This step leads to sessile chrysophytes with golden or off-green chloroplasts. For a more complete account of these organisms see Bourrelly (1968) or Starmach (1985). *Poterioochromonas* is attached to the substrate by a delicate secreted lorica; *Ochrostylon* by a thin thread drawn out from the posterior of the body.

Chrysophytes include some species with chloroplasts, and some without. The group gets its name from the golden colour of the photosynthetic pigments. If plastids are present, they frequently occur singly or in pairs (Fig. 23). There is usually a small stigma or eyespot in one plastid, although it may be very difficult to see. Many species, including those with plastids, can ingest particles of food, i.e. they are mixotrophic (see Sanders in Patterson and Larsen, 1991). Chrysophytes include species that are colonial (e.g. Figs 24 & 46), solitary (e.g. Figs 22 & 28), with plastids (Figs 23 & 52) or without (Figs 47 & 101). They are related to diatoms (Figs 6–9), oomycete fungi, bicosoecids (Fig. 32) and brown algae (see Green *et al.*, 1989).

Chrysophytes have a long and a short (sometimes absent) flagellum: the long flagellum bears stiff hairs which are invisible by light microscopy. The action of this flagellum draws a current of water towards the body surface. Most species are small (5–10 μm long). Many form flask-shaped siliceous cysts with a small pore for egress (stomatocysts: Fig. 24). Some swimming species (e.g. of *Paraphysomonas*, Figs 28 & 101) temporarily attach to the substrate.

Guides to genera and species may be found in Bourrelly (1968), Starmach (1980), and Patterson and Larsen (1991). The general biology is discussed by Green *et al.* (1989) and in Patterson and Larsen (1991).

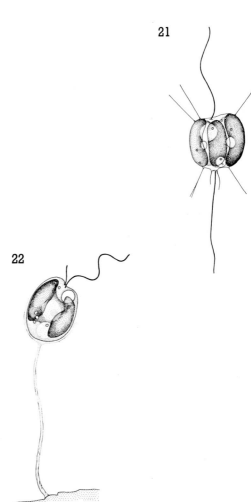

21

22

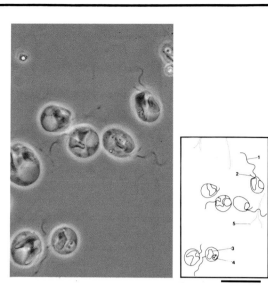

Figure 23 *Poterioochromonas.* A chrysophyte (chryso-monad), most of which are, like this species, very small. The cells have two flagella which insert near the apex. One flagellum is relatively long (1), but the other is much shorter (2) and tends to flop back over the body. The beating of the flagellum draws a current of water along it towards the cell. The cell contains a single, curved, golden or orange (chryso = golden) chloroplast (3). Most species in this genus are mixotrophic, eating bacteria as well as carrying out photosynthesis; the presence of food vacuoles containing bacteria (4) is evidence of this. *Poterioochromonas* is very like *Ochromonas*, but differs in its ability to form a very delicate, long, stalked, eggcup-like lorica (5), in which the cells usually sit. *Phase contrast.*

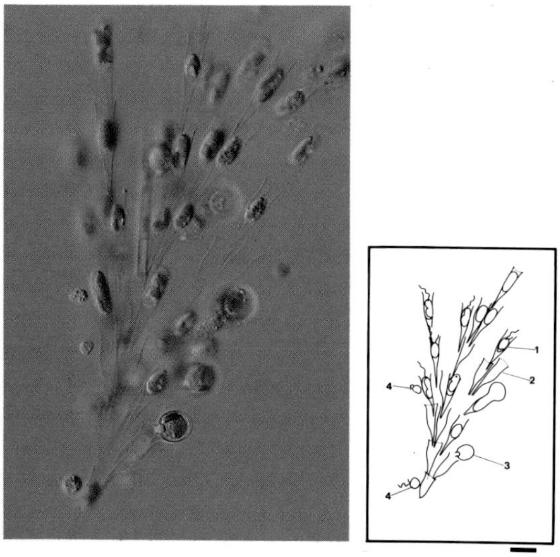

Figure 24 *Dinobryon.* A colonial chrysophyte. Each cell has two flagella, of which one is short and one long (1). Near the anterior margin of this cell is a dark, refractile droplet, the stigma or eyespot. The cells have golden chloroplasts and live within vase-shaped organic lorica (2), attached to each other to give the fan shape to the colonies. When conditions are no longer ideal, the cells encyst within a siliceous stomatocyst (3), which has a plugged aperture for excystment. Attached to this colony are a number of small colourless chrysophytes (4). *Differential interference contrast.*

7
(5)

A   Cells without evident covering.

GO TO 8

B   Cells mostly enclosed with a covering (lorica or test).

GO TO 10

8
(7)

A   Cells that attach to the substrate by one flagellum, and are easily distinguished by their characteristic flicking movements when attached. They may detach and swim. 5–10 µm long.

**Figs 25 & 69 BODO SALTANS
(PLEUROMONAS JACULANS)**

Bodo saltans (= *Pleuromonas jaculans*) is a bodonid (Step 27). There is some debate as to which is the correct name for this organism. *B. saltans*, which has been described in detail by Brooker (1971), feeds on suspended bacteria, using the shorter anterior flagellum. The attachment structure may be identified as a flagellum because it is motile, and because careful scrutiny will reveal that it is inserted along-side the second shorter flagellum.

25

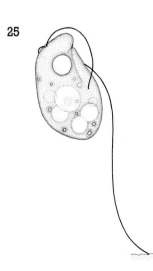

B   Cells not attached by flagella and without the kicking movement.   GO TO 9

A   Cells with one flagellum at the unattached end, a delicate cytoplasmic stalk for attachment to the substrate, and a number of fine arms extending from the cell.   Figs 26 & 27 ACTINOMONAS

Actinomonas is the most common colourless pedi-nellid (Step 6) found in fresh water. The arms may be withdrawn under some circumstances and the cells may detach to swim in wide arcs. Members of the genus *Actinomonas* may have arms projecting from all parts of the cell (Larsen, 1985). The body measures 5–20 µm in diameter. A second genus, *Pteridomonas*, has arms only around the flagellum, and is more common in marine sites (Larsen and Patterson, 1990; Patterson and Fenchel, 1985). *Para-physomonas vestita* (Figs 28 & 101) is a chrysophyte coated with a layer of spines. It may sometimes attach to the substrate, and can be distinguished from pedinellids because it has two flagella of un-equal length.

26

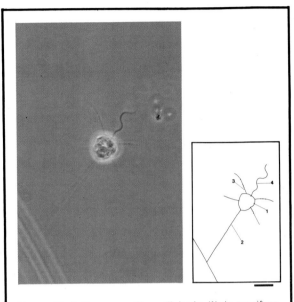

Figure 27 *Actinomonas*. The cell body (1) is pomiform (apple-shaped). From the posterior end emerges a stalk (2) with which the cell may attach temporarily to the substrate. More usually, it will swim in lazy circles, with the stalk trailing behind. A single flagellum (4), beating with a planar sine wave, emerges at the anterior end of the cell. Around this project stiff arms (3) which bear small granules (ex-trusomes). The cells are filter feeders, drawing a current of water through the arms. Particles are then trapped against the arms, probably by secretions released from the ex-trusomes. The arms are withdrawn when the cells are swimming. *Phase contrast.*

ALL SCALE BARS 20 µm UNLESS OTHERWISE INDICATED

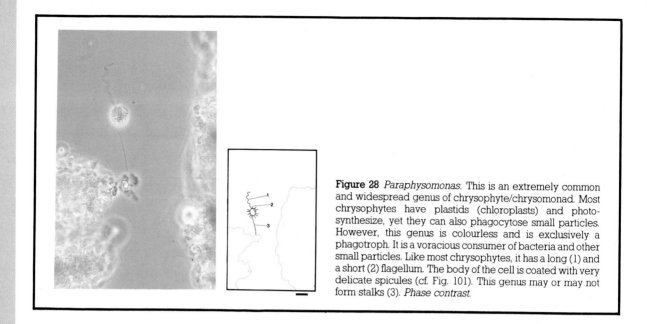

Figure 28 *Paraphysomonas.* This is an extremely common and widespread genus of chrysophyte/chrysomonad. Most chrysophytes have plastids (chloroplasts) and photosynthesize, yet they can also phagocytose small particles. However, this genus is colourless and is exclusively a phagotroph. It is a voracious consumer of bacteria and other small particles. Like most chrysophytes, it has a long (1) and a short (2) flagellum. The body of the cell is coated with very delicate spicules (cf. Fig. 101). This genus may or may not form stalks (3). *Phase contrast.*

B   Cells with a single apical flagellum surrounded by a fine cytoplasmic collar (collar flagellates). The flagellum draws a current of water through the collar from its base towards its apex. Mostly under 10 μm long.

Figs 29(a) & (b) & 30 MONOSIGA

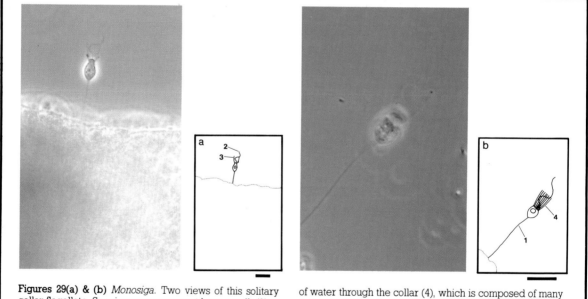

Figures 29(a) & (b) *Monosiga.* Two views of this solitary collar flagellate. Species may or may not have a stalk (1) at one apex, with which they attach to the substrate. At the other apex of the cell is a single flagellum (2) which beats with a sine wave in a single plane. The beat draws a current of water through the collar (4), which is composed of many fine cytoplasmic fingers. The individual fingers are usually not visible, and the collar is normally only seen in profile, as two extensions (3), one on either side of the flagellum. *Phase contrast.*

Collar flagellates are common in freshwater and marine environments. The cells may be naked, or invested in an organic lorica or a siliceous case (marine species only). The cytoplasmic part of the cell is similar in most species, with a body (5–10 μm in diameter) giving rise to a single flagellum and the collar of pseudopodia. Genera and species are distinguished by the form of the lorica, by being free-swimming or attached, by being stalked or un-stalked, or by being solitary or colonial (see Patterson and Larsen, 1991; Zhukov and Karpov, 1985).

Almost all species are attached, although abrasion can detach them, or, if conditions become unfavourable, they may release themselves and swim around with the flagellum directed backwards.

The collar traps bacteria and is made of very fine pseudopodia (Fig. 29(b)) which are normally seen only as two lines, one on either side of the flagellum (Fig. 29(a)). Trapped bacteria are drawn into the cell by a pseudopodium which extends from the cell body. The biology of the group is reviewed in Patterson and Larsen (1991), and guides to species may be found in Zhukov and Karpov (1985), Bourrelly (1968), Ellis (1929), and Starmach (1980). More detailed descriptions are provided by Leadbeater and Morton (1974), Hibberd (1975) and Andersen (1989).

Compare with *Paraphysomonas* (Fig. 28).

A Cells with a fine cytoplasmic collar around the single flagellum (see notes after Step 9). The lorica is organic, forming either a thin, transparent sheath or a thick, brown shell around the cell.   GO TO 11

10
(7)

B Cells without a cytoplasmic collar, living in a vase-shaped lorica, into which they may suddenly retract. These are two flagella: one emerges beside a slight lip at the top of the cell, and is held in a gentle curve; the other bends sharply backwards to attach to the base of the lorica. Cells may be clustered to form pseudocolonies. Typically, the bodies are small (5–10 μm long).   Figs 31 & 32 BICOSOECA

30

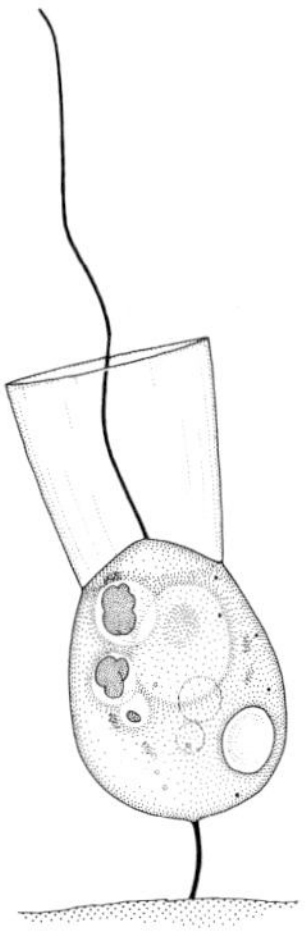

31

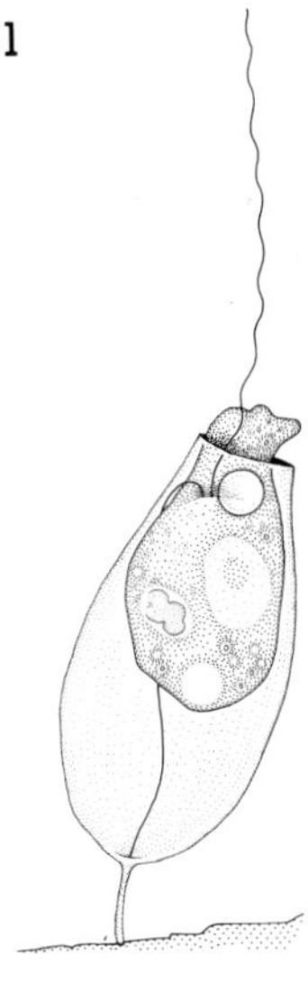

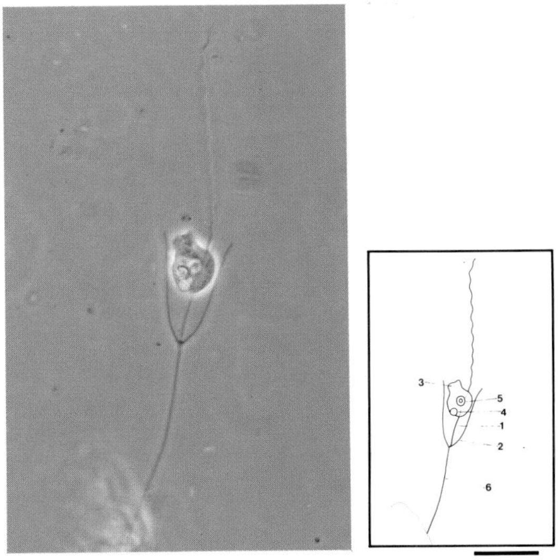

Figure 32 *Bicosoeca*. A genus of colourless, filter-feeding flagellates. The cells have no plastids, and are immotile and loricated. Mostly attached to immersed surfaces, but colonial forms may be found suspended in the water column. There are two flagella, both inserting near the top of the cell. One (1) curves backwards to attach to the base of the lorica (2). With suitable stimuli, this flagellum will 'contract' to pull the cell into the lorica. The other flagellum is long, beats with a shallow wave, and draws a current of water to the cell surface, where particles carried in the current are intercepted by a projecting lip (3). A contractile vacuole (4), food vacuoles (5) and the lorica stalk (6) are visible. *Phase contrast.*

Species of *Bicosoeca* (often misspelt *Bicoeca*) are colourless flagellates related to the chrysophytes. They are filter feeders, using the long flagellum to draw a current of water to the body surface. Particles (bacteria) impinge near the anterior lip or shoulder, where they are then ingested. The recurrent flagellum can contract to pull the cell back into the lorica, at which point both flagella become coiled.

There is only one genus in fresh water, but two naked genera, *Pseudobodo* and *Cafeteria*, are widespread in marine environments (Larsen and Patterson, 1990; Patterson and Larsen, 1991). Detailed accounts of fine structure are provided by Mignot (1974) and Moestrup and Thomsen (1976). The cells' general biology is described by Picken (1941), feeding behaviour by Sleigh (1964), and identities of species by Zhukov (1978). Planktonic species are discussed by Hilliard (1971).

**11
(10)**

A Collar flagellates with a thin colourless, organic lorica that may closely adhere to the surface of the body. Only the anterior portion may be visible, like a second collar around the flagellum. Body diameter is about 5–10μm. See notes after Step 9.   Fig. 33 SALPINGOECA

B Small flagellates living in a thick lorica that becomes brown with age and encrusts on the substrate to form volcano-like mounds. The collar and the flagellum can be seen, usually with great difficulty, protruding from an anterior pore. Body diameter is about 5 μm. See notes after Step 9.   Figs 34 & 35 DIPLOECA

The taxonomy of collar flagellates with thick loricae that become brown is confused. Two genera, *Diploeca* and *Pachysoeca*, were erected by Ellis (1929), but *Salpingoeca* and *Diplosigopsis* house similar species (Bourrelly, 1968; Starmach, 1980). The bacterium *Siderocapsa* also forms brown, volcano-like deposits.

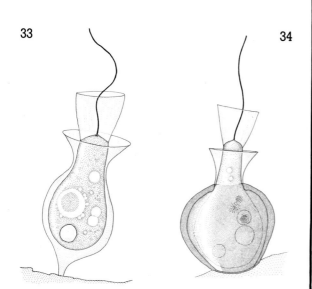

33   34

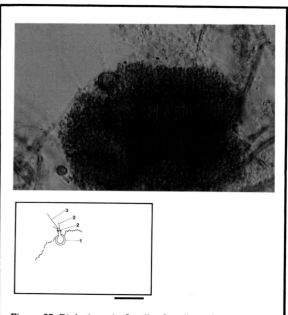

Figure 35 *Diplosigopsis*. A collar flagellate. One of several genera with a brown, thickened test (1). This genus can be distinguished from the others (e.g. *Pachysoeca* and *Diploeca*) because there appear to be two cytoplasmic collars (2) around the apical flagellum (3). Often only the test is visible, in which case, confusion with the bacterium *Siderocapsa* is possible. This particular cell is attached to a mucilaginous mass produced by *Spongomonas* (Fig. 40), and the long orange rods are iron-impregnated secretions of the iron bacterium, *Galionella*. *Differential interference contrast.*

A   Colourless cells.   GO TO 13

B   Elongate cells (mostly about 10 μm long) living in a branching colony of vase-shaped loricae (10–100 μm long) that attach to each other. Each cell has two flagella, only one of which is easy to see. The cells have golden chloroplasts and a stigma.   Figs 24 & 36 DINOBRYON

Dinobryon is a type of mixotrophic chrysophyte (see notes after Step 6). Many of the common species are described by Bourrelly (1968), and some aspects of the fine structure have been described by Owen *et al.* (1990a). *Dinobryon* is often encountered as motile planktonic colonies.

36

A   Cells embedded in mucus.   GO TO 14

B   Cells not in mucus; usually single or in masses at the ends of narrow branches.   GO TO 16

A   Fan-shaped colonies in which mucus forms flattened and grooved sheets. The cells are small (about 5–10 μm in diameter) and located in tubes at the ends of the mucus sheets, and each cell has two flagella.   Fig. 37 RHIPIDODENDRON

Previously regarded as a chrysophyte (e.g. Starmach, 1980), but detailed studies (Hibberd 1976c) indicate otherwise. One species is found in *Sphagnum* moss.

37

B   Mucus in bulbous masses and with a globular consistency.   GO TO 15

A Pear-shaped cells with a single flagellum that emerges through a stiffened collar at the cell apex. Colonies usually in hemispherical masses. Individual cells about 10 μm long. **Fig. 38 PHALANSTERIUM**

Only a few species are known, of which one is said to be solitary. However, these cells may simply be 'seeds' for new colonies. The fine structure has been described by Hibberd (1983).

38

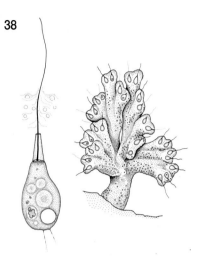

B Small, rounded cells (body about 10 μm long) with two flagella. The flagella beat stiffly, and sometimes have a very shallow basal collar. Colonies may measure hundreds of microns.

Figs 39 & 40 SPONGOMONAS

Colonies are hemispherical or finger-shaped, sometimes branching (Schneider, 1986), and often have a pinkish or brownish colour. The fine structure has been described by Hibberd (1976c, 1983).

39

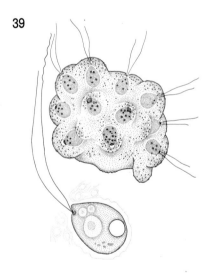

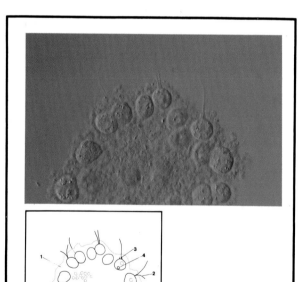

Figure 40 *Spongomonas*. A colourless filter-feeding flagellate that occurs in bulbous gelatinous colonies. The jelly-like matrix of the colony has the texture of adhering globules. Each cell has two projecting flagella, sometimes with a shallow collar around the bases (3). A contractile vacuole (4) lies near the base of the cell. *Differential interference contrast.*

A Cells in clusters at the ends of narrow branches of the colony.   GO TO 17

B Each branch of the colony terminates in a single cell (body 5–10 μm).

Figs 41 & 42(a) & (b) PSEUDODENDROMONAS

The cells are somewhat triangular in profile, and have two equally long flagella at one 'corner'. The stalks are fairly wide, often with mucus and accumulated debris around them. Ultrastructure is described by Mignot (1974b) and Hibberd (1985). There is a similar chrysophyte (see notes after Step 6), *Dendromonas*, which forms branching colonies of cells borne on stalks. The two genera can be distinguished from each other as the cells of *Dendromonas* have thin stalks and the flagella are different in length. *Pseudodendromonas* is related to *Cyathobodo* (not illustrated) which is solitary, but may also secrete a stalk.

41
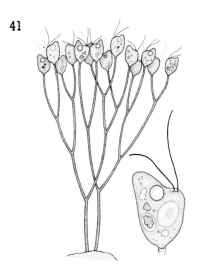

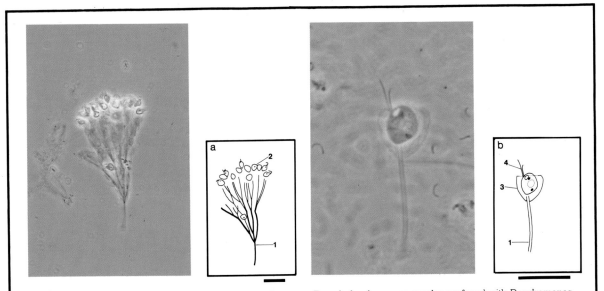

Figures 42(a) & (b) *Pseudodendromonas*. A colourless filter-feeding flagellate that typically occurs in a fan-shaped colony. The cells are borne on a branching stalk system (1) to which organic matter and bacteria may adhere. The cells lie in an arc at the head of the colony (2). Each cell is held within a vase-shaped lorica (3) and has two flagella (4).

Pseudodendromonas may be confused with *Dendromonas*, each cell of which also has two flagella (but they are very unequal in length), or with *Rhipidodendron* (see Fig. 37), the cells of which are supported in a globular, fluted, organic matrix. *Phase contrast*.

A   The flagellum of each cell is surrounded by a fine cytoplasmic collar (collar flagellates: see notes Step 9). The body is 5–10 μm in diameter, and the stalk is firm.

Figs 43 & 44 CODOSIGA

One species is described in detail by Hibberd (1975), and Leadbeater and Morton (1974). The genus has also been referred to as *Codonosiga*. For species taxonomy see notes after Step 9.

43

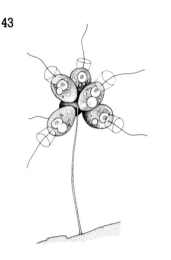

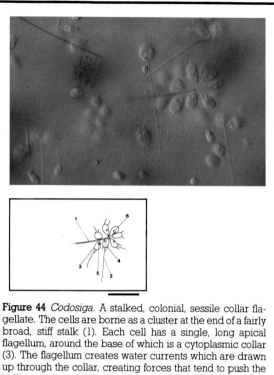

Figure 44 *Codosiga*. A stalked, colonial, sessile collar flagellate. The cells are borne as a cluster at the end of a fairly broad, stiff stalk (1). Each cell has a single, long apical flagellum, around the base of which is a cytoplasmic collar (3). The flagellum creates water currents which are drawn up through the collar, creating forces that tend to push the cell towards the substrate. The stalk is thick to prevent compaction (cf. *Paraphysomonas*, Fig. 28). Nuclei with a central nucleolus (4), food vacuoles (5) and empty-looking, basal contractile vacuoles (6) are evident. *Differential interference contrast.*

B   No collar around the flagellum/a.

GO TO 18

ALL SCALE BARS 20 μm UNLESS OTHERWISE INDICATED

A Cells occurring in clusters at the ends of softish, dichotomously dividing branches with a granular consistency. The material of the stalks becomes brown with age, being darker near the base of the stalks and virtually colourless near the cells. The cell bodies measure 5–10 μm in diameter, and there are two unequal flagella. The colonies may attach to debris, vegetable matter or to the air–water interface.

Figs 45–47 **ANTHOPHYSA**

A. vegetans is described in detail by Pringsheim (1946) and by Belcher and Swale (1972). The species is a chrysophyte (see notes after Step 6).

45

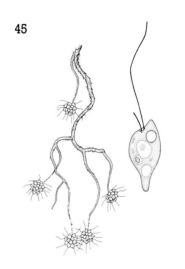

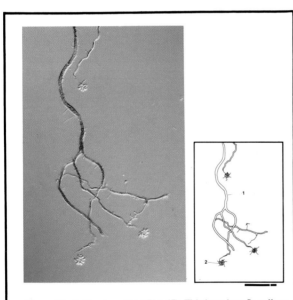

Figure 46 *Anthophysa* (see Fig. 47). This is an iron flagellate and a chrysophyte. It takes the form of branching (arborescent) colonies, usually attached at the broad end to the substrate, or hanging from the water–air interface. They are called iron flagellates because the organic matter of the stalk (1) accumulates metal salts, such as iron and manganese, giving them their rusty colour. The cells are arranged in small, spherical clusters (2) at the ends of each of the branches. They can occur in large numbers, turning surfaces brown. *Differential interference contrast.* Scale bar 100μm.

Figure 47 *Anthophysa* (see Fig. 46). Illustrated is a single cluster of cells that has begun the process of producing a stalk. The youngest part (1) of the stalk is adjacent to the colony; it is lightest in colour because it has absorbed only a small quantity of metal salts. The cells are drawn out where they attach to the stalk. At the opposite end, each cell has two unequal flagella (as do all chrysophytes), emerging from a slight dimple. *Differential interference contrast.*

B The stalk is rigid, colourless and unbranching. Cells 7–15 μm long.   Fig. 48 CEPHALOTHAMNIUM

A colonial bodonid (see notes after Step 27), described in detail by Hitchen (1974), each cell is attached to the common stalk by one recurrent flagellum. The cells may occur as epizoites.

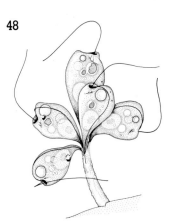

48

19
(3)

A Cells forming colonies.   GO TO 20

B Cells not forming colonies.   GO TO 26

20
(19)

A The colony is a spherical ball of colourless cells, held together in a mass of mucus. Each cell has a single flagellum, around which projects a fine cytoplasmic collar (collar flagellates, see notes after Step 9 above). The individual cells are small (body 5–10 μm in diameter).   Figs 49 & 50 SPHAEROECA

Studies of colonial collar flagellates have been conducted by Leadbeater (1983) and Ertl (1981).

49

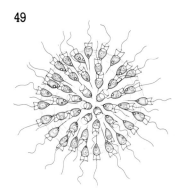

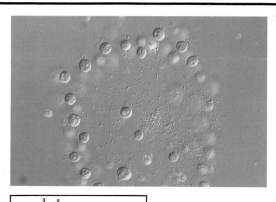

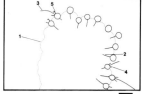

Figure 50 *Sphaeroeca*. A planktonic colonial collar flagellate. All of the cells lie at the outer edge of a ball made of gelatinous material (1). The colony may measure as much as 300 μm in diameter. Each cell has its posterior end (2) drawn out into the matrix, and has a single flagellum (3) which is surrounded by a collar (4). Bacteria from the feeding current adhere to the collar (5) before ingestion. *Differential interference contrast*. (Photo Helge Thomsen.)

B The cells of the colony contain chloroplasts, so the colony has a green or golden colour. GO TO 21

A The colony is golden (colonial chrysophytes: see notes after Step 6). Most take the form of spherical colonies, although *Dinobryon* colonies (Figs 24 & 36) are feather-shaped. GO TO 22

Most of the planktonic colonial chrysophytes have siliceous scales (i.e. they are members of Synurophyceae = synurophytes). Species and genera with scales (and spines) usually have to be studied by electron microscopy before they can be identi- fied (see Andersen (1986a, b), Kristiansen (1975), Kristiansen and Andersen (1985), Moestrup and Andersen in Patterson and Larsen (1991), and Wee (1982).

B The cells of the colony contain bright green chloroplasts (i.e. with chlorophyll b). The colonial volvocids. GO TO 24

Solitary or colonial motile members of green algae are here referred to as volvocids, in accordance with the protozoological literature (e.g. Lee *et al.*, 1985). The phycological literature may differ. The number of cells in each colony usually helps to iden- tify organisms to genus. The number ranges from four to thousands, and the size of the colony from 20 µm to over a millimetre. Colony colour ranges from a pastel to a deep green. Each cell is rigid, as a cellulose cell wall is present. The cells are usually embedded in mucus through which flagella, usually in pairs, protrude.

The volvocids are an ecologically successful group of green algae. The cells typically bear two or, less commonly, four flagella. Flagella of one cell are of the same length, beat with a breaststroke action, and may adhere to the substrate. The cells typically have a single, bright green chloroplast, with a stigma and pyrenoid within it. Land plants are related to this group. The group contains many genera of solitary species (e.g. *Chlamydomonas*, Figs 113 & 114) as well as genera of colonial organisms. Identification guides by Bourrelly (1972), Pentecost (1984), and Prescott (1978). One or two genera lack chloroplasts (Figs 93–95).

A Cells (5–15 µm) without scales and in gelatinous colonies up to 150 µm in diameter.
Figs 51 & 52 UROGLENA

Uroglena cells are most easily confused with *Synura* (Step 23B), and care is needed to discern the scales of the latter genus. They may also be confused with the less common *Syncrypta*, but in that genus the colonies are smaller, the cells are densely packed, and the flagella are virtually equal in length, (not markedly unequal). *Uroglena* cells are connected by thread-like extensions of the posterior end. As with most chrysophytes, they have two flagella of unequal length, two chloroplasts, and a stigma lying outside the chloroplast. Studies have been conduc- ted by Owen *et al.*, 1990b (see also Hibberd and Leedale (1985), and notes after Step 6).

51

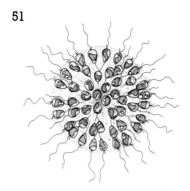

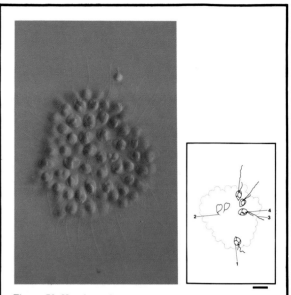

Figure 52 *Uroglena*. A spherical, swimming colonial chrys-ophyte. The form of the colony resembles that of some colonial green algae (e.g. Fig. 61). The individual cells are embedded in the outer regions of a mucilaginous material (not visible here). Related genera have the cells joined together. The individual cells have golden chloroplasts (1) and two flagella of unequal length. (cf. Fig. 53). *Differential interference contrast.*

B Cells with scales and/or spines, not obviously embedded in mucus.   GO TO 23

23
(22)

A Each cell with a coating of scales and spines. Cells 5–20 μm long.   Fig. 53 CHRYSOSPHAERELLA

B Each cell with scales only, cells normally 15–40 μm long.   Figs 54 & 55 SYNURA

Detailed accounts of the fine structure of *Synura* are given by Andersen (1985), Brugerolle and Bricheux (1984), and Schnepf and Deichgraber (1969); and of *Chrysophaerella* by Asmund (1973), Nicholls (1980) and Andersen (1990). For identification by light microscopy see Bourrelly (1968), and for diagnosis by electron microscopy of scales see Starmach (1985).

53

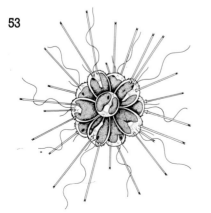

54

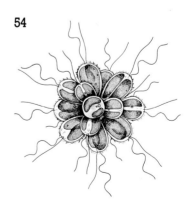

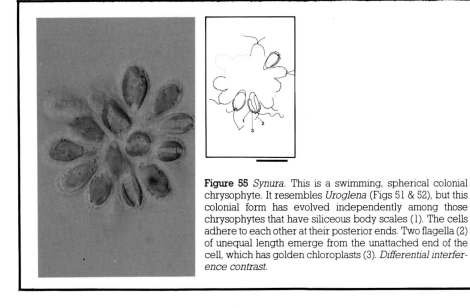

Figure 55 *Synura*. This is a swimming, spherical colonial chrysophyte. It resembles *Uroglena* (Figs 51 & 52), but this colonial form has evolved independently among those chrysophytes that have siliceous body scales (1). The cells adhere to each other at their posterior ends. Two flagella (2) of unequal length emerge from the unattached end of the cell, which has golden chloroplasts (3). *Differential interference contrast.*

A The form of the colony is a flat plate. Cells are 5–15 μm in diameter, and colonies are up to 100 μm.

Fig. 56 GONIUM

24
(21)

For taxonomy see notes to Steps 21 & 25.

56

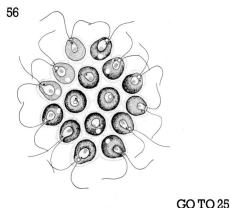

B The colony is spherical.

GO TO 25

A Each cell is relatively large (up to 20 μm) in relation to the size of the colony (compare Figs 56–61) and they actually or nearly touch at their posterior ends. Usually eight or 16 cells in a colony.

Fig. 57 PANDORINA

25
(24)

B Numerous relatively small cells forming a colony which is like a hollow ball, within which more densely packed daughter colonies may be seen. Cells are usually less than 10 μm in diameter, but colonies can be larger than 1 mm.

Figs 58–60 VOLVOX

There are several other genera of colonial volvocids: *Platydorina* colonies are flat and less regular than *Gonium*. *Eudorina* (Fig. 61) is like *Pandorina* (Fig. 57), but the cells are usually more numerous (32) and do not touch. There are a number of other genera that form spherical colonies. They vary with respect to the number, size and arrangement of cells, the number of flagella, and the shape of the colony. Fuller accounts are given by Bourrelly (1972) and Ettl (1983).

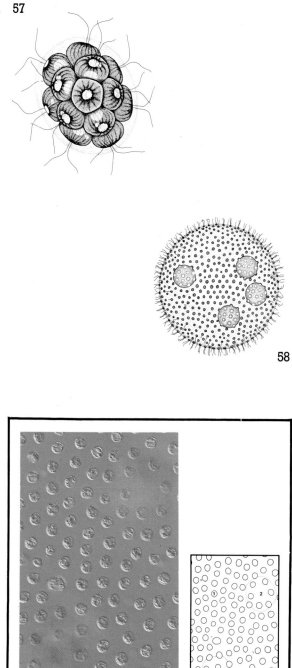

58

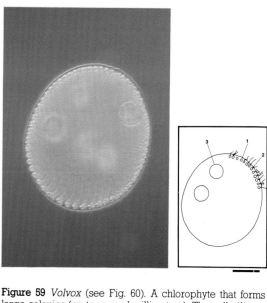

Figure 59 *Volvox* (see Fig. 60). A chlorophyte that forms large colonies (up to several millimetres). The cells (1) are embedded in a gelatinous matrix, from which project the flagella (2) that propel the cell. Daughter colonies develop within the parental colony, and can be seen as brighter green inclusions (3). They break free by rupturing the surface of the colony. *Differential interference contrast.* Scale bar 100 μm.

Figure 60 *Volvox.* A detail of the surface of the colony (see Fig. 59). The individual cells (1) embedded in the gelatinous matrix may be seen; behind them are the slightly darker spheres of daughter colonies (2). Each cell has a chloroplast. Flagella are not visible. *Differential interference contrast.*

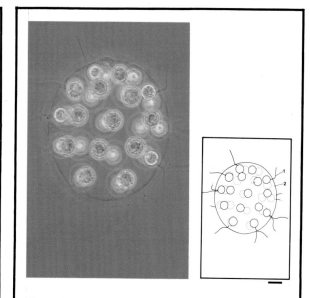

Figure 61 *Eudorina.* A colonial, motile chlorophyte. One of the most evident adaptive traits within the swimming chlorophytes has been the evolution of motile colonies. Such colonies may take the form of flat plates of cells (*Gonium*), a tightly packed cluster of cells (*Pandorina*) or, as here, cells more loosely aggregated within a gelatinous matrix. Each cell has two flagella (1) which project through the matrix (2). The cells are arranged in five circumferential bands: two bands of four cells each, and three of eight cells. *Phase contrast.*

A   The cells lack chloroplasts.   GO TO 27

B   The cells have chloroplasts.   GO TO 54

A   Cells whose normal movement is a smooth gliding in close contact with the substrate.   GO TO 28

Mostly colourless euglenids and bodonids. The flagellum/a trail/s against the ground and one (at least) is relatively immotile. The body may be very plastic or even amoeboid.

B   Cells that normally swim rather than glide.   GO TO 39

Care is needed to establish the 'normal' mode of locomotion. Most species that normally glide may lose contact with the substrate and then begin to swim. Preparations should be left for several minutes for such cells to settle. Some flagellates that normally swim may come to rest against debris in order to feed: a small number of volvocids (see Step 59) lay their flagella against a solid substrate and glide using them (Bloodgood, 1981). Care should be taken to distinguish gliding from squirming, which may result when cells are trapped in debris or between the slide and the coverslip.

A   Small cells (less than 10 μm) with a single trailing flagellum and what appears to be a vibrating bulbous 'nose'.   Figs 62 & 63 RHYNCHOMONAS

The 'nose' contains a cytostome (mouth) and is supported by a short flagellum. It is pressed against individual bacteria before they are ingested. Described in detail by Swale (1973) and Burzell (1973), *Rhynchomonas* is easily confused with another gliding flagellate referred to as *Amastigomonas* (Patterson and Larsen, 1991) or *Thecamonas* (Larsen and Patterson, 1990), which has a parallel-sided snout and a trailing flagellum that is rarely seen.

    Rhynchomonas is a bodonid flagellate and is closely related to trypanosomes (both being Kine-

62

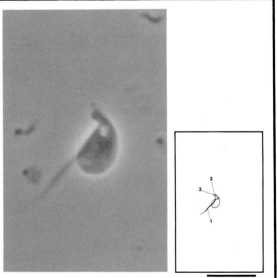

Figure 63 *Rhynchomonas*. A common bodonid flagellate. It is not typical of the group as only one flagellum (1) is obvious. The (posterior) flagellum is thicker near the cell body than it is at the tip because of a paraxial rod that lies alongside the axoneme in the anterior part. This is a common feature of many bodonids, and may be used as a 'rule-of-thumb' to identify members of this difficult group. The second flagellum, of which a small portion may be seen here (2), supports the snout. The snout (3) contains the cytostome and wobbles from side to side as the cell moves across the substrate. *Phase contrast*.

toplastids). Unlike trypanosomes, bodonids have two flagella and most are free-living. Bodonids are typically very small (rarely larger than 15 µm), and species in particular are extremely difficult to distinguish. Most bodonids glide or skip across the substrate with one flagellum trailing rather inactively behind the cell (but see *Bodo saltans* Step 8). The posterior part of the trailing flagellum is thinner (acronematic), and the contractile vacuole is located in the anterior part of the cell, normally near the discrete mouth and the anterior insertion of the flagella. The mouth is usually used to prise individual bacteria from the substrate.

Identifying a small (less than 15 µm) flagellate as a bodonid is rarely easy, as the diagnostic feature, the kinetoplast (a mass of DNA in the mitochondrion), cannot normally be seen in living cells. There are only two common genera (*Bodo* and *Rhynchomonas*). Many small flagellates have arbitrarily been grouped with the bodonids. Confusion is possible with small euglenids (Figs 82–84); euglenids can usually be distinguished because the posterior flagellum is rarely prominent and the anterior flagellum is rather thick. Cercomonads (Step 31) may also cause confusion, but they can normally be distinguished by their readiness to form pseudopodia, because the trailing flagellum adheres to the body surface, and because the contractile vacuole is often located in the posterior part of the cell. For detailed accounts of the cytology of bodonids see Vickerman and Preston (1976), Brooker (1971), Brugerolle *et al.* (1979) and Patterson and Larsen (1991). Species descriptions are given by Hänel (1979), Vickerman (1976) and Zhukov (1971).

B   Cells with one or two typical flagella and no other appendages.   GO TO 29

29
(28)

A   Cells with one or both flagella lying along the substrate as the cell moves.   GO TO 30

B   Cells with one, two or four flagella at the apex of the cell, connected to the nucleus. The flagella beat stiffly and the cell is flexible.   Figs 20 & 64 FLAGELLATED CELL OF MYXOGASTREID SLIME MOULD
Figs 85 & 86 MASTIGAMOEBAE

The flagellated swarmers of myxogastreid (myxomycete) slime moulds (Olive, 1975) are not common in freshwater, but they may be encountered in samples and cultures from soils or vegetation. They may have one flagellum visible, although usually two or more may be seen. Typical sizes are 10–20 µm.

Mastigamoebae (Step 39) have a single, stiff flagellum, with the base either attached directly to the nucleus (*Mastigamoeba*) or removed from it (*Mastigella*). They may be confused with myxomycete slime moulds, but they rarely glide, and are most often encountered in sites lacking oxygen and usually smelling of hydrogen sulphide.

64

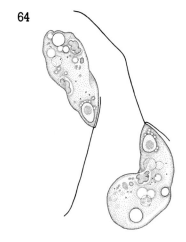

30
(29)

A   Cells with two projecting flagella.   GO TO 31

B   Cells with one projecting flagellum.   GO TO 37

A The cells are very flexible and sometimes amoeboid. The anterior flagellum has a stiff sweeping motion, and the posterior flagellum adheres to the body surface near its insertion. The cells are usually, but not always, less than 15 μm in diameter, with very pliant bodies from which pseudopodia may emerge. The body may be drawn out along trailing flagellum, or as strands from the posterior end of the cell. The posterior flagellum may be thicker for its proximal (first) half.   Figs 65 & 66 CERCOMONAS

65

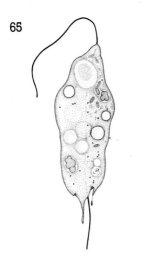

Figure 66 *Cercomonas*. A gliding flagellate with a very pliable body, one anterior beating flagellum (1) and a posterior trailing flagellum (2) which adheres to the cell body (at least at its most anterior part). The cell cytoplasm is often drawn out behind the cell as it moves. *Cercomonas* is mostly bacterivorous, preferring bacteria attached to surfaces. *Phase contrast.*

Sometimes referred to as *Cercobodo*, this cell is not a bodonid, but is related to the common soil flagellate *Heteromita* and a marine flagellate *Massisteria* (Patterson and Larsen, 1991). Cercomonads tend to form pseudopodia when feeding, a feature that helps distinguish them from bodonids (Steps 28 & 32), and the cytoplasm may contain granules. The cell bodies have small extrusomes. (Mignot and Brugerolle, 1975; Schuster and Pollak, 1978) Some confusion may also occur with the amoeboid *Mastigamoeba* and *Mastigella*, both of which have a single flagellum (Step 39, Fig. 86).

B The cells are not amoeboid, but they have a distinctive shape, and they may or may not writhe.

GO TO 32

A   Small cells, which generally measure less than 15 µm, with two flagella that are inserted to one side of the most anterior part of the cell, or along one side (not ventrally). The flagella are not markedly thickened, and the anterior one is more active. The body may either be firm or plastic, but if the cells are uncompressed, they do not writhe. They eat small particles.

Figs 67–69 BODO

See notes after Step 27.

67

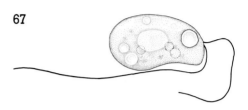

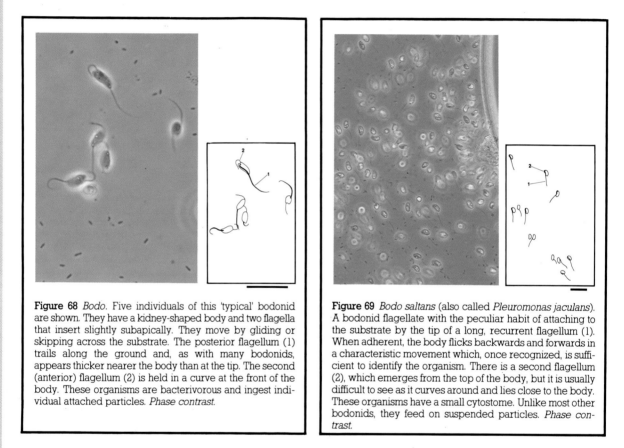

Figure 68 *Bodo*. Five individuals of this 'typical' bodonid are shown. They have a kidney-shaped body and two flagella that insert slightly subapically. They move by gliding or skipping across the substrate. The posterior flagellum (1) trails along the ground and, as with many bodonids, appears thicker nearer the body than at the tip. The second (anterior) flagellum (2) is held in a curve at the front of the body. These organisms are bacterivorous and ingest individual attached particles. *Phase contrast.*

Figure 69 *Bodo saltans* (also called *Pleuromonas jaculans*). A bodonid flagellate with the peculiar habit of attaching to the substrate by the tip of a long, recurrent flagellum (1). When adherent, the body flicks backwards and forwards in a characteristic movement which, once recognized, is sufficient to identify the organism. There is a second flagellum (2), which emerges from the top of the body, but it is usually difficult to see as it curves around and lies close to the body. These organisms have a small cytostome. Unlike most other bodonids, they feed on suspended particles. *Phase contrast.*

B   Cells in which one or both flagella appear unusually stout, and the body surface may be helically striated or twisted. The flagella insert in a small depression or pocket which opens near the anterior pole of the cell or slightly ventrally. Smaller species are usually rigid, while larger species can often writhe. These cells are rarely less than 10 µm.

THE GLIDING EUGLENIDS   GO TO 33

Euglenids are a well known group (Leedale, 1967; Patterson and Larsen, 1991), the taxonomy of which is described by Huber-Pestalozzi (1955), Bourrelly (1970), Buetow (1982), and Patterson and Larsen (1991). Best known through the swimming species that have chloroplasts with chlorophyll b (e.g. *Euglena*, Figs 108, 119–121), the group contains many colourless genera that are heterotrophs. Some of these swim and are osmotrophic, i.e. they absorb soluble nutrients (e.g. *Astasia*, Fig. 89), but most glide and often have an ingestion apparatus (e.g. *Entosiphon*, Figs 75 & 76) with which they ingest particles of food. Despite the variety of ways in which they obtain food and energy, euglenids have many ultrastructural features in common. The body surface is fluted, folded or grooved because of underlying skeletal strips (sometimes visible with the light microscope, Figs 120 & 121), and most larger phagotrophic species can writhe (metaboly or euglenoid motion) (Suzaki and Williamson, 1986a & b). Euglenids are related to bodonids, and the smallest gliding euglenids occupy ecological niches similar to those of bodonids.

A   Cell body is not rigid.   GO TO 34

33
(32)

B   Cell body is rigid.   GO TO 35

A   Cells with a prominent anterior flagellum, and a small ingestion apparatus (body 20–100 μm long).
                                                                              Figs 70–72 PERANEMA

34
(33)

Peranema (botanists call this *Pseudoperanema*, see Patterson and Larsen, 1991) squirms actively, especially when feeding. Ingestion of food involves the use of two ingestion rods lying near the anterior tapering pole of the cell (Fig. 71; Nisbet, 1974). The cytoplasm is often heavily laden with starch granules, and the pellicle is finely ridged. The front flagellum is very strongly developed, with most movement occurring near the tip. Extremely careful observation is needed to detect the second flagellum, which lies in a slight groove in the ventral surface of the body (Fig. 72).

70

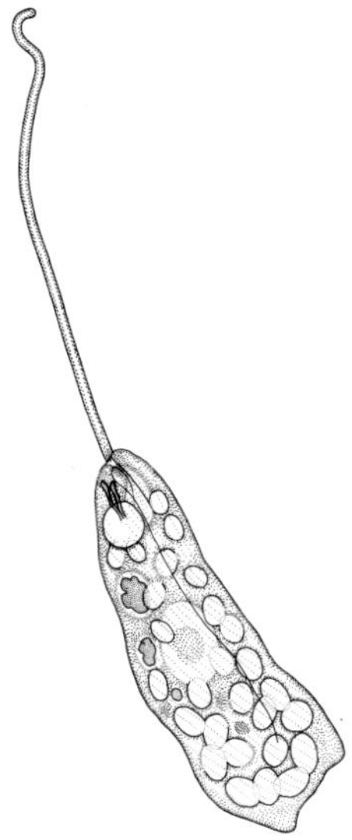

ALL SCALE BARS 20 μm UNLESS OTHERWISE INDICATED

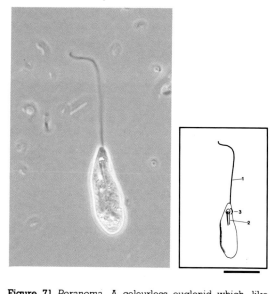

Figure 71 *Peranema*. A colourless euglenid which, like many other colourless euglenids, does not readily swim, but glides along the ground. It appears to have only a single, very broad, emergent flagellum (1). During normal locomotion, the basal part remains fairly stiff, with only the anterior portion showing much activity. *Peranema* (also called *Pseudoperanema*) is a phagotroph and can manipulate other protists and detritus into the cell by means of two ingestion rods (2). This species does also have a short recurrent flagellum (3). *Phase contrast.*

Figure 72 *Peranema*. A detailed view of the anterior end of this colourless and phagotrophic euglenid (cf. Fig. 71, but note that this is a larger species). Although only one flagellum (1) appears to emerge from the front of the cell, careful scrutiny reveals a second recurrent flagellum (2) leaving the reservoir to extend backwards, lying close to the surface of the cell. Both flagella remain in close contact with the substrate (in this picture they are in the same focal plane as the bacteria adhering to the substrate (3)) as the cell moves. Delicate striations of the pellicle (4), typical of many euglenids, can also be seen. *Differential interference contrast.*

B   Cells with an anterior flagellum, a trailing flagellum that is not attached to the body surface and can be seen easily, and a small ingestion device. Size 20–200 μm.   Figs 73 & 74 HETERONEMA

73

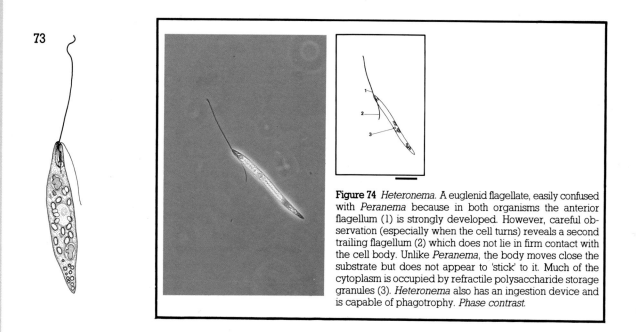

Figure 74 *Heteronema*. A euglenid flagellate, easily confused with *Peranema* because in both organisms the anterior flagellum (1) is strongly developed. However, careful observation (especially when the cell turns) reveals a second trailing flagellum (2) which does not lie in firm contact with the cell body. Unlike *Peranema*, the body moves close the substrate but does not appear to 'stick' to it. Much of the cytoplasm is occupied by refractile polysaccharide storage granules (3). *Heteronema* also has an ingestion device and is capable of phagotrophy. *Phase contrast.*

A Cells without ingestion apparatus. GO TO 36

B Cells with a well-developed ingestion apparatus (body 20–30 μm long). Figs 75 & 76 ENTOSIPHON

These cells have an anterior beating flagellum and a second, broad, trailing flagellum. Some ultrastructural aspects are presented by Mignot (1966) and Triemer and Fritz (1987). Two species are common in freshwaters. In marine sites a rather similar genus, *Ploeotia*, is common (Larsen and Patterson, 1990).

75

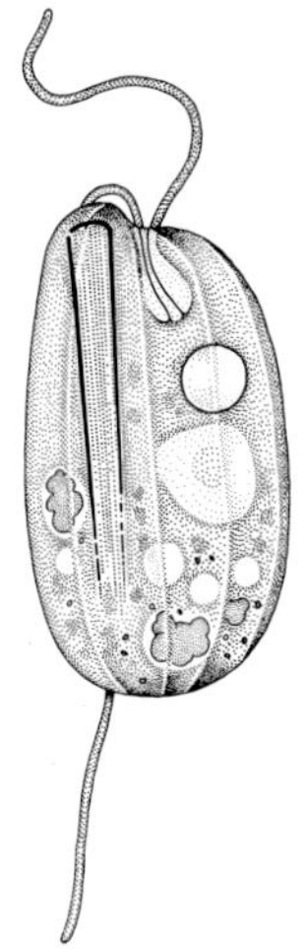

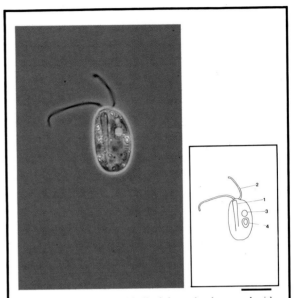

Figure 76 *Entosiphon*. Of all of the colourless euglenids, *Entosiphon* has the most strongly developed ingestion apparatus (1). This comprises a tube with a flap-like opening at its anterior end. Detritus and bacteria are ingested through this organelle. There are two flagella (2), but this photograph is misleading because one normally trails behind the cell as it moves, while the other (the anterior) beats in a fairly conventional fashion. The contractile vacuole (3) and nucleus (4) are also evident within the cell. *Phase contrast*.

A The posterior flagellum is very broad at its base, and is as long as or longer than the anterior one. The forward motion of the cell is occasionally interrupted by backwards jerks (body 10–100 μm long).
Figs 77 & 78 ANISONEMA

The posterior flagellum curves in a broad arc like a 'hook' as it leaves the flagellar pocket. The occasional jerks in motion are caused by contraction of the posterior flagellum. For taxonomy, see Huber-Pestalozzi (1955) and Larsen and Patterson (1990).

Helkesimastix (Fig. 81) has two flagella and may key out here, although the anterior flagellum cannot usually be seen and it appears to be uniflagellated.

77

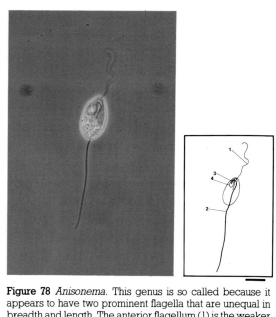

Figure 78 *Anisonema*. This genus is so called because it appears to have two prominent flagella that are unequal in breadth and length. The anterior flagellum (1) is the weaker of the two, beating normally. The recurrent flagellum (2) is much broader and trails along the ground as the cell moves. Morphologically, the clearest distinguishing feature is the 'hook' (3) that the recurrent flagellum forms after leaving the flagellar pocket (4). Living cells can also be distinguished from other genera because the recurrent flagellum can contract to jerk the cell backwards. No ingestion apparatus is visible. *Phase contrast.*

B The posterior flagellum is not greatly thickened, and is shorter than the anterior one. Most species are 15–30 μm long.   Figs 79 & 80 NOTOSOLENUS

79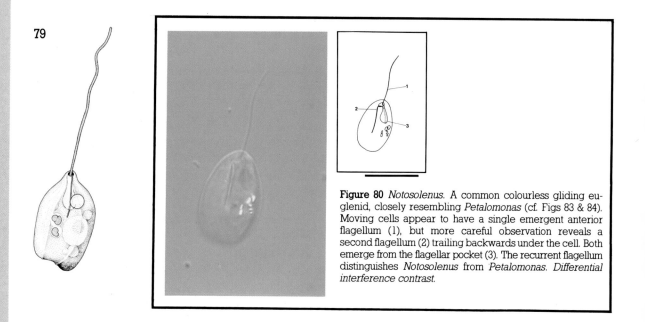

Figure 80 *Notosolenus*. A common colourless gliding euglenid, closely resembling *Petalomonas* (cf. Figs 83 & 84). Moving cells appear to have a single emergent anterior flagellum (1), but more careful observation reveals a second flagellum (2) trailing backwards under the cell. Both emerge from the flagellar pocket (3). The recurrent flagellum distinguishes *Notosolenus* from *Petalomonas*. *Differential interference contrast.*

Careful scrutiny is needed to see the trailing flagellum. It is usually most visible if the cell turns. Confusion with *Petalomonas* (Step 38), which does not have a trailing flagellum, is likely. *Notosolenus* cells are often slightly flared at the anterior end. In *Notosolenus* and *Petalomonas*, the anterior flagella are held stiffly in front of the cell and motion is mostly confined to its anterior portion. Cells are often wedge-shaped and the body surface may bear ridges or be fluted. Ingestion devices cannot usually be seen with the light microscope, but the cells may contain large particles of food, which indicate that a mouth is present (Larsen and Patterson, 1990).

A Cells with the single flagellum directed anteriorly during normal locomotion. Mostly euglenids.

GO TO 38

See notes on euglenids after Step 32.

B Cells with the single flagellum directed posteriorly. Body less than 10 μm long.

Fig. 81 HELKESIMASTIX

This genus has a tiny recurrent flagellum which is readily overlooked. This and several similar taxa (e.g. *Allas* and *Allantion*) from soils (see Sandon, 1927; Patterson and Larsen, 1991) have been reported rarely, and may be cercomonads with reduced anterior flagella.

81

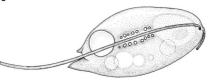

A Rigid cells. Most species are 10–50 μm long.

Figs 82–84 PETALOMONAS

82

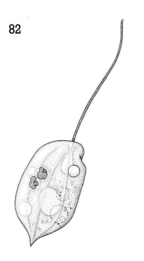

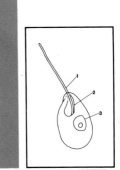

Petalomonas is generally a small cell (under 15 μm) in which the flagellum is most active at the anterior tip. It is most easily confused with *Notosolenus* (Step 36) which has a rather insignificant recurrent flagellum. Taxonomy is discussed in Huber-Pestalozzi (1955), Shawan and Jahn (1947), and Larsen and Patterson (1990).

Figure 83 *Petalomonas*. A colourless euglenid flagellate with a single emergent flagellum only (1), arising from a flagellar pocket (2). As the cell glides, the flagellum lies along the substrate, apparently pulling the cell forwards. The anterior end of the flagellum is its most active part. No ingestion organelle is visible in cells of this genus, at least by light microscopy. Ultrastructural studies suggest that small ingestion devices may be present. The nucleus (3) is atypically obvious in this cell. *Phase contrast.*

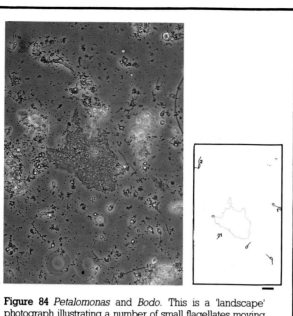

Figure 84 *Petalomonas* and *Bodo*. This is a 'landscape' photograph illustrating a number of small flagellates moving among bacteria. The bacteria include coccoid, filamentous and spiral forms. Two types of flagellate are present: *Petalomonas* (cf. Fig. 83) (1), which includes smaller species of euglenids, and is distinguishable by the single stiff anterior flagellum; and two bodonids (2), each of which has two flagella. Euglenids and bodonids are thought to be related; small species like these occupy similar niches, being associated with detritus and consuming small attached particles. *Phase contrast.*

ALL SCALE BARS 20 μm UNLESS OTHERWISE INDICATED

B **Highly metabolic cells.** **GO TO 39**

39 (40) **A** **Thick flagellum, most active near the tip. The body is capable of deformation, but is not amoeboid.** **GO TO 40**

B **Almost amoeboid body. The flagellum is thin and beats stiffly, like an undulating rod. Body 10–100 μm long.** **Fig. 85 MASTIGAMOEBA**

The relationships of *Mastigamoeba* have only recently become clear. Related to *Mastigella* (Fig. 86) and *Pelomyxa*, the body resembles that of an amoeba. The flagellum of *Mastigamoeba* is attached to the nucleus, whereas that of *Mastigella* is removed somewhat from the nucleus. *Pelomyxa* looks like an amoeboid organism, the flagella being relatively short and insignificant. The flagellum is rather stiff and flexes rather than undulates. Some mastigamoebae are very similar to swarmers from certain slime moulds (Fig. 20). Reviewed in Lemmermann (1914) and Patterson and Larsen (1991).

85

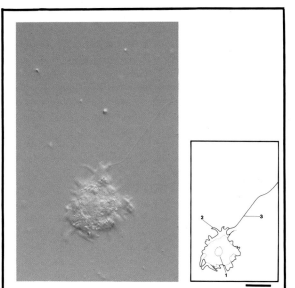

Figure 86 *Mastigella*. An organism that combines the characteristics of an amoeba and of a flagellate, i.e., it has an amoeboid body and a flagellum. The mastigamoebae are not well known, but two genera, *Mastigamoeba* and *Mastigella*, are reasonably common. *Mastigamoeba* has its nucleus lying at the base of the flagellum, whereas in *Mastigella* it (1) lies near the centre of the cell. Pseudopodia (2) develop from the body surface. There is a single, long flagellum (3) which beats very stiffly (rather like a flexing stiff rod). There is great similarity with the 'swarmers' of some slime moulds (Fig. 20). The mastigamoebae are usually found in organically enriched or anoxic sites. *Differential interference contrast.*

A Cell tapers at the front end. Most species 20–100 μm long.   Figs 70–72 PERANEMA   **40 (39)**

See Step 34. *Peranema* has two flagella, but the recurrent one is extremely difficult to see. *Pera-* *nemopsis* includes virtually identical organisms, but they do not have the recurrent flagellum.

B Cell flares at the front end. Body 25–60 μm long.   Figs 87 & 88 URCEOLUS

Urceolus has an ingestion apparatus comprised of two rods to manipulate food. The stout anterior flagellum resembles that of *Peranema* in its behaviour, being most active at the tip. The surface of the cell may be finely ridged and, in some species, particles adhere to the surface. Can be fairly large (up to 50 μm).

87

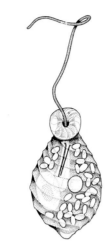

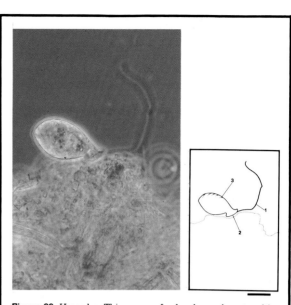

Figure 88 *Urceolus*. This genus of colourless, phagotrophic euglenid flagellates has much in common with *Peranema* (Figs 71 & 416). The body is highly metabolic and there is a single, broad, emergent flagellum (1) which extends in front of the cell as it glides along, but the genus is distinguished by the flared anterior end (2). Surface striations, characteristic of many euglenids, are visible (3). *Urceolus* also has a rod-like ingestion device (not visible) and eats detritus, algae and other protists. *Phase contrast.*

41
(27)

A With one or two groups of one to four flagella.   GO TO 42

B With more than four beating flagella, but not emerging in groups.

There are several rarely encountered organisms that satisfy this description. They include *Artodiscus* (see Rainer (1968) where it is regarded as an amoeba), *Multicilia*, *Spironema*, *Psalteriomonas* and *Hemimastix* (Broers *et al.*, 1990; Foissner *et al.*, 1988; Patterson and Larsen, 1991).

42
(41)

A With one flagellum.   GO TO 43

B With two or more flagella.   GO TO 46

Care must be taken here because in some organisms with more than one flagellum, only one flagellum is readily visible.

43
(42)

A Flagellum directed forwards.   GO TO 44

B Flagellum directed backwards.

Very few protozoa satisfy this description. However, dislodged collar flagellate cells (see Step 9B) will swim with their flagellum trailing behind. The same is true of some fungal swarmers, particularly those chytrids (Fig. 8) with a long, trailing flagellum and a small, spherical body containing a refractile granule. Dinoflagellates (Step 67), most of which are pigmented, have two flagella, but often only the trailing flagellum is seen.

A Relatively broad flagellum, beating in whiplash fashion (i.e. small coils are pushed along the flagellum from base to tip). Most species are 20–80 μm long.
Figs 89–92 COLOURLESS EUGLENIDS
e.g. ASTASIA, MENOIDIUM

Astasia (Figs 89 & 92) is one of several genera of actively swimming colourless euglenids (see notes after Step 32). As with all euglenids, there is an anterior depression or pocket from which two flagella arise, although usually only one emerges. Some genera, like *Astasia*, are highly metabolic; others, such as *Menoidium* (Figs 90 & 91), are relatively rigid. They are often found in organically polluted sites, duck ponds, etc. Ultrastructure has been described by Suzaki and Williamson (1986b).

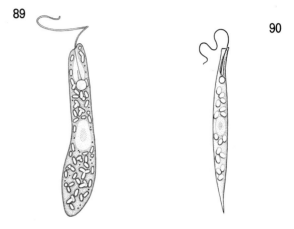

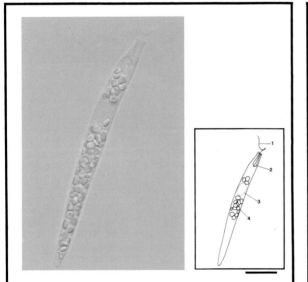

Figure 91 *Menoidium*. A swimming colourless euglenid flagellate without an ingestion organelle, *Menoidium* appears to survive using some form of osmotrophic nutrition (absorbing soluble nutrients from the medium). A single emergent flagellum (1) arises in the flagellar pocket (2) which lies slightly behind the anterior pole of the cell. The nucleus (3) has a slightly punctate appearance. Note the loop in the flagellum, illustrating a type of flagellar (whiplash) beating encountered only in euglenids. Much of the cytoplasm is occupied by polysaccharide storage granules (4). *Differential interference contrast.*

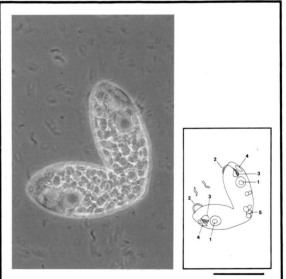

Figure 92 *Astasia* dividing (longitudinal division). Division in most flagellates begins with duplication of the flagellar structures, followed by mitosis, and then by the progression of the division furrow from the front to the back of the cell. In this cell the nuclei (1) have divided. The single emergent flagellum (2) from each cell is short, and inserts into the flagellar pocket (3). The contractile vacuoles (4) release their contents into the flagellar pocket. *Phase contrast.*

B Thin flagellum, beating not as a whiplash but more in a breaststroke or undulating fashion.   GO TO 45

45
(44)

A Thin flagellum, beating in a planar sine wave. Usually trailing a thin stalk or strand of cytoplasm. The body is small (usually less than 10 μm) and apple-shaped.

(Step 9) DETACHED ACTINOMONAD FLAGELLATES

B Flagellum inserting at the conical pole of an almost amoeboid cell, beating rather like a flexing rod. Cells small, usually under 20 μm.   (see notes after Step 39) SLIME MOULD SWARMER

46
(42)

A Two or four flagella, equal in length, beating with a breaststroke movement at the apex of the cell. The cells are usually ovoid or have blunt posterior protrusions. Most species are 10–30 μm long.

Figs 93–95 COLOURLESS VOLVOCIDS

There are two common genera: *Polytomella* (Fig. 93) with four flagella (de la Cruz and Gittelson, 1981), and *Polytoma* (Fig. 94) with two flagella and a cellulose wall surrounding the cell. For a general guide to volvocid literature, see notes after Step 21. For colourless genera, see Pringsheim (1937), Lang (1967) and Gaffal and Schneider (1980).

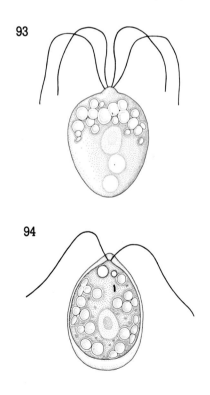

93

94

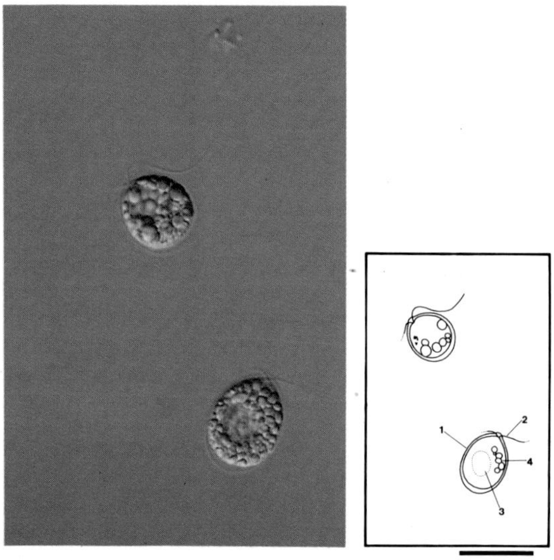

Figure 95 *Polytoma*. A motile, coccoid chlorophyte (cf. Fig. 113), this is one of the few colourless genera. The cells have no chloroplasts, but retain many of the other distinguishing features of chlorophytes. The body is enclosed within an organic cell wall (1), and there are two apical flagella (2). (*Polytomella* is a related genus with four flagella and no cell wall). The nucleus (3) lies near the centre of the cell, and the cytoplasm typically appears very granular because of polysaccharide storage materials (4). These granules (often referred to as starch) are refractile and may appear to have a greenish tinge under some lighting conditions. Care should be taken to confirm whether green pigment is present or not. The simplest means of doing this is to view the cell with bright-field optics and with the condenser iris fully open. *Differential interference contrast.*

B Not as 46A.   GO TO 47

A   Inflexible body that tapers posteriorly or is sigmoid, with two flagella emerging together from an anterior groove or channel that is surrounded by an aggregation of small refractile bodies. The cell body is rounded in cross section and typically contains numerous refractile starch grains. Cells swim freely or come to rest near detritus. Normally 20–40 μm long.   Figs 96 & 97 CHILOMONAS

Cryptomonads are common. Most genera contain off-green, blue-green, golden or reddish chloroplasts, and may occur in blooms (natural occurrences of high densities of cells). Generic identification of organisms with plastids usually requires electron microscopy (Patterson and Larsen, 1991). There are two colourless genera, *Chilomonas* being particularly widespread and a weed. The other genus is *Goniomonas* (Step 48). Whether with or without chloroplasts, most cryptomonads have bodies that are rounded or only slightly flattened in cross section, with two flagella emerging from the anterior opening of a groove (often misleadingly referred to as a gullet). The body usually tapers and twists slightly posteriorly. There is normally one contractile vacuole per cell: this vacuole discharges into the flagellar pocket.

The cell sometimes has large, pinkish refractile crystals. The refractile bodies around the flagellar depression are extrusible organelles called ejectisomes. They are expelled by trapped or otherwise distressed cells, causing them to jump suddenly backwards. Other behaviour includes forwards swimming (flagella divergent but directed forwards), backwards swimming (which may cause difficulties in identifying the front and the back of the cell), and resting (little flagellar action). The ultrastructure of *Chilomonas* is described by Roberts (1981b). *Kathablepharis* (p. 181) is a colourless flagellate with two lines of refractile bodies. It is sometimes allied with the cryptomonads (Bourrelly, 1970) and may be the same as the marine *Leucocryptos* (Patterson and Larsen, 1991). For identification at the light-microscopical level, see Bourrelly (1970); for marine species, see Butcher (1967).

96

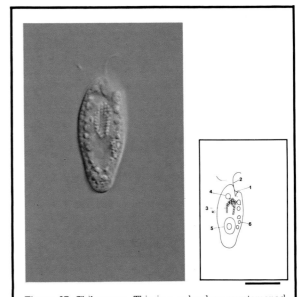

Figure 97 *Chilomonas.* This is a colourless cryptomonad (cf. Fig. 126), but it has a typical cryptomonad shape: a rigid body, often with the posterior end narrowed (sometimes even pointed). The anterior end of the cell is indented (1) where the flagellar pocket or groove opens. Two flagella (2) project from the groove which, inside the cell, is lined by extrusible ejectisomes (3). The contractile vacuole (4) lies near the most anterior shoulder of the cell. The nucleus (5) is relatively large and much of the cytoplasm is filled with 'starch' grains (6). *Differential interference contrast.*

B   Not as 47A.   GO TO 48

48
(47)

A Small (5–10 µm) flattened cells, with two divergent flagella arising together near an anterior lateral corner of the cell. Movement is by skidding parallel to the substrate. A single refractile bar runs parallel to the anterior margin of each cell. **Figs 98 & 99 GONIOMONAS**

Goniomonas, often called *Cyathomonas* (for name change see Larsen and Patterson, 1990), is an atypically shaped member of the cryptomonads (Step 47). Ultrastructure is described by Mignot (1965) and in Patterson and Larsen (1991).

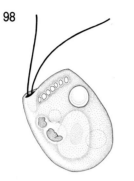

98

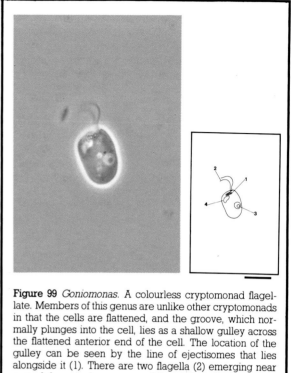

Figure 99 *Goniomonas*. A colourless cryptomonad flagellate. Members of this genus are unlike other cryptomonads in that the cells are flattened, and the groove, which normally plunges into the cell, lies as a shallow gulley across the flattened anterior end of the cell. The location of the gulley can be seen by the line of ejectisomes that lies alongside it (1). There are two flagella (2) emerging near one of the anterior corners of the cell, a single median nucleus (3), and an anterior contractile vacuole (4). This genus is common, although it is rarely reported. It normally moves by skidding along the substrate, and is usually known as *Cyathomonas*. *Phase contrast.*

B Cells with unequal flagella or with flagella not emerging at the same point on the cell surface. **GO TO 49**

49
(47)

A Cells with a long, undulating flagellum held in a gentle arc extending from the front of the cell, and a second short flagellum that curves backwards to lie near the cell surface. Colourless chrysophytes (see Step 6). **GO TO 50**

B Cells with two or more flagella. If there are two, they are equal in length. The flagella may emerge from opposing sides of the cell. **GO TO 51**

50
(49)

A Cells coated in a layer of very delicate spicules. Body 5–20 µm. **Figs 100 & 101 PARAPHYSOMONAS**

Paraphysomonas is a colourless chrysophyte (Step 6). The spicules may be evident only as a halo around the cell. This is a very common genus, the cells of which may swim around or temporarily attach to the substrate, either by using a thread-like extension of the posterior end of the cell, or by secreting a delicate mucoid stalk (Figs 28 & 100). The scales of most species in the genus are too small to be visible with the light microscope, and generic and species identification requires electron microscopy (Preisig and Hibberd, 1982, 1983a, 1983b; Vørs *et al.*, 1990).

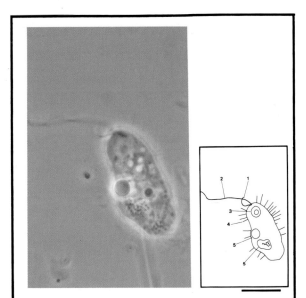

100

Figure 101 *Paraphysomonas.* A colourless chrysophyte with the one short flagellum (1) and one long (2) flagellum that are characteristic of this group. The nucleus (3) lies near the bases of the flagella. The genus is distinguished by having no chloroplasts (the golden object is a food vacuole) and by being coated with a layer of delicate spicules (4). The scales of most species in this genus can only be seen by electron microscopy. Inside the body lie numerous food vacuoles (5) with different kinds of ingesta. *Phase contrast.*

B   Cells without spicules. Body 5–20 μm long.

Spumella is a colourless chrysophyte (Step 6). As members of this genus are identical to many species of *Paraphysomonas* (Step 50A) when viewed with the light microscope, the absence of scales must be confirmed by electron microscopy. The ill-defined genus *Monas* is regarded as being the same as *Spumella* (see Preisig *et al.* in Patterson and Larsen, 1991). Detailed descriptions are given by Mignot (1977), and taxonomy is discussed by Bourrelly (1967) and Starmach (1985). Individuals may attach temporarily to the substrate and may eat bacteria or other small protists.

Fig. 102 SPUMELLA

102

A   With flagella emerging from opposing sides of the cell, laterally, posteriorly or anteriorly. Mostly under 20 μm long.   **FREE-LIVING DIPLOMONADS   GO TO 52**

Most diplomonads are parasites (Lee *et al.*, 1985; Patterson and Larsen, 1991), but a small number of genera occur in natural bodies of water, usually in organically polluted sites and under fairly anoxic conditions. They typically possess two nuclei and clusters of four flagella which arise at the anterior ends of lateral grooves in the body. Genera and species differ in the relative length (and therefore the visibility) of the flagella. Some species swim and turn with a characteristic stepwise rotation. For general comments, see Patterson and Larsen (1991); for descriptions of free-living species, see Calaway and Lackey (1962), Hänel (1979), and Lemmermann (1914). Electron microscopy is discussed by Eyden and Vickerman (1975) and by Brugerolle in Patterson and Larsen (1991).

B   With flagella arising together, at, or near, the apex of the cell.   **GO TO 53**

A   With four flagella on either side of the body. One is long and projects laterally, while the other three are shorter and difficult to see. Cell body 7–30 μm long.   Figs 103(a) & 105 **TREPOMONAS**

B   The flagella not only extend laterally, but they may also lie in the groove from the point of flagellar insertion, and trail behind the cell, or even project in front of it. Cell mostly 10–30 μm long.   Figs 103(b), 104 & 106 **HEXAMITA**

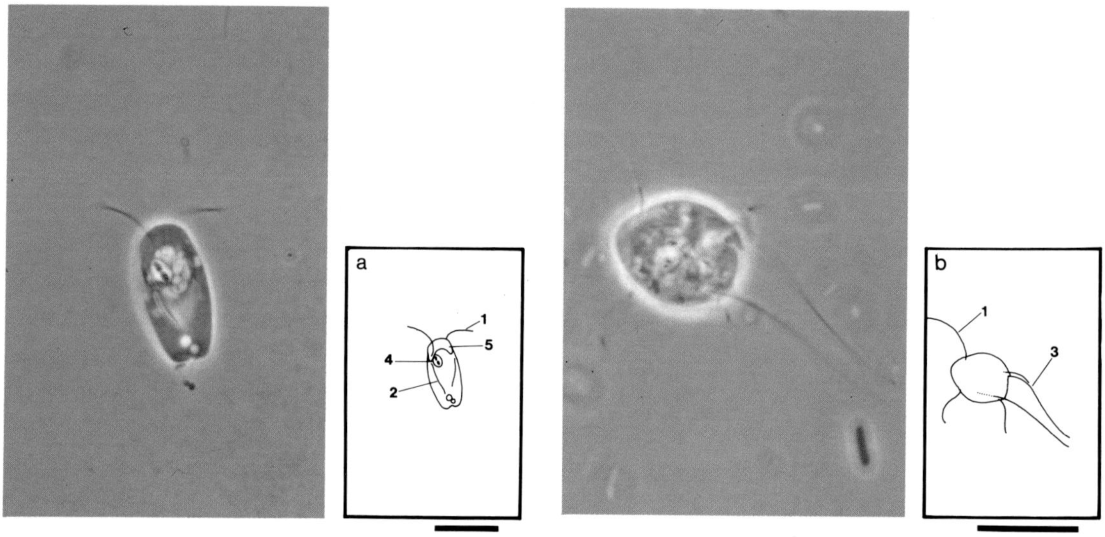

Figures 103(a) & (b) Diplomonad flagellates, *Trepomonas* (a) and *Hexamita* (b). Most genera of diplomonads are parasites, and the few genera that are free-living are usually found in organically enriched (and usually anaerobic) sites. The cells are bilaterally symmetrical along their longitudinal axis. There are two anterior nuclei (5), and associated with each are four flagella which arise at the head of a groove in the body surface. The genera may be distinguished by the relative lengths of the flagella and by the flexibility of the bodies. In both genera, one flagellum of both quartets extends laterally (1) from the head of the groove. The remainder lie within the groove, with those of the more pliable *Trepomonas* (2) not extending beyond the posterior margin of the cell, as do those of *Hexamita* (3). These organisms may feed either by eating bacteria (4) or by pinocytosis. *Phase contrast.*

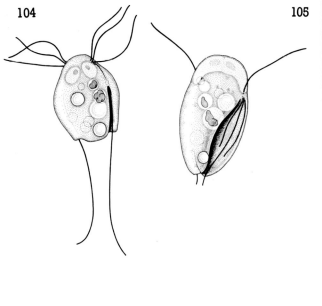

104   105

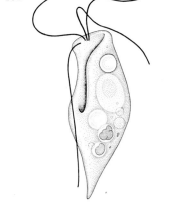

Figure 106 *Hexamita.* A diplomonad. In this genus, one flagellum projects forwards (1), and the remainder lie in the groove. One of the latter may be seen on each side at the posterior end of the cell (2). The cells have food vacuoles containing bacteria (3). *Phase contrast.*

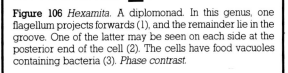

A   With four flagella and a slit extending from the site of flagellar insertion. 10–20 μm long.

Fig. 107 TETRAMITUS

53
(51)

B   No slit and two flagella. Cell bodies 10–20 μm long.

(Step 112) NAEGLERIA

Tetramitus and *Naegleria* (Fig. 206) are hetero-loboseids (Page and Blanton, 1985; Patterson and Larsen, 1991). In both genera the flagellate is one stage of a polymorphic life cycle that also involves amoebae and cysts. For light microscopy of *Tetramitus*, see Bunting (1926) and Bunting and Wenrich (1929), and for ultrastructure, see Balamuth *et al.* (1983). *Naegleria* is of interest since one free-living species is able to invade the central nervous system through the nasal mucosa, and causes a fatal meningitis. This species is found in warm waters (Martinez, 1985).

Slime mould swarmers (Figs 20 & 64) may have two or more flagella and may key out here. Flagella insert at the apex of the cell, with a cone holding the nucleus near to the flagellar bases.

107

A   The chloroplasts (and cell) are bright green (chlorophyll b present).

GO TO 55

54
(39)

B   The colour is off-green, golden or red.

GO TO 62

55
(54)

A The cell is rigid, with a smooth organic wall and two or four flagella of equal length, beating with breaststroke action.   (Step 21) VOLVOCIDS   GO TO 56

B With one thick flagellum, beating with a whiplash motion (coils are pushed along the flagellum from base to tip). The cell may squirm or, if rigid, it is usually spirally sculpted. One genus has a round lorica from which a single, long flagellum emerges.   EUGLENIDS   GO TO 60

Euglenids and volvocids are the only types of flagellate to have bright green (grass green) chloroplasts (compare the types of plastid in Fig. 108). Members of the two groups can be distinguished fairly easily as euglenids normally have one emergent thick flagellum and can squirm, while volvocids have two or more thin flagella and are rigid. With the exception of *Trachelomonas* (Step 63), euglenids do not have surrounding cell walls. Both types of flagellate contain a stigma, but this is located within a chloroplast in volvocids, and in the cytoplasm of euglenids. Euglenids have a flagellar pocket and a nucleus usually with a granular consistency. Some colourless euglenids have been keyed out already (see Step 32 and the following steps) as have colonial and colourless volvocids (Step 24, the following steps, and Step 46). Only a small number of genera of volvocids are keyed out below. For a guide to the appropriate literature, see notes after Step 21.

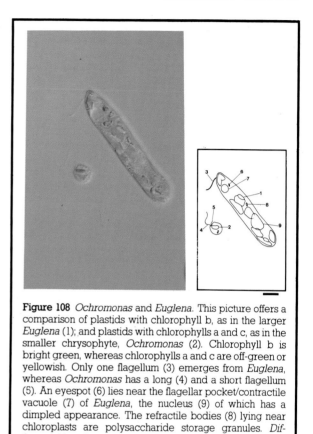

Figure 108 *Ochromonas* and *Euglena*. This picture offers a comparison of plastids with chlorophyll b, as in the larger *Euglena* (1); and plastids with chlorophylls a and c, as in the smaller chrysophyte, *Ochromonas* (2). Chlorophyll b is bright green, whereas chlorophylls a and c are off-green or yellowish. Only one flagellum (3) emerges from *Euglena*, whereas *Ochromonas* has a long (4) and a short flagellum (5). An eyespot (6) lies near the flagellar pocket/contractile vacuole (7) of *Euglena*, the nucleus (9) of which has a dimpled appearance. The refractile bodies (8) lying near chloroplasts are polysaccharide storage granules. *Differential interference contrast.*

56
(55)

A The cells are spindle-shaped, with the flagella located at one apex. Cells 20–200 μm, mostly about 30 μm.   Fig. 109 CHLOROGONIUM

109

B The cells are not spindle-shaped. GO TO 57

A The wall is pressed close to the surface of the cell. GO TO 58

B The cells attach to the wall by means of thin strands of cytoplasm. Cell body 20–70 μm.

Figs 110 & 111 (a) & (b) **HAEMATOCOCCUS**

Haematococcus can develop a bright red pigment that masks the green colour. Consequently it reappears in this key as a non-green flagellate (Step 62). This adaptive red coloration is encountered in some euglenids, and is held to be a protection against intense radiation. *Haematococcus* is often found in shallow puddles, where it stains the water red. Joyon (1965) gives information on ultrastructure.

110

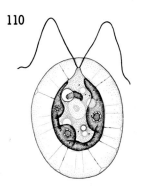

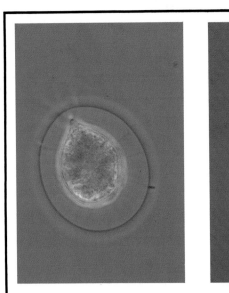

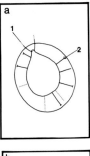

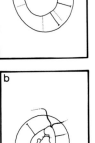

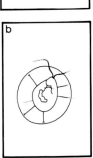

Figures 111(a) & (b) *Haematococcus*. A motile chlorophyte. Each cell has two flagella (1) and is enclosed in a stiff organic theca, to which it is attached by thin strands of cytoplasm (2). The cells have chlorophyll b in their chloroplasts, usually giving them a bright green colour. However, under some circumstances (e.g. intense radiation) they will develop an additional scarlet pigment which turns the cell red. This genus occurs in shallow puddles, and the development of the second pigment in dense blooms of cells may cause the water to turn red, or cause a red film to develop on the substrate and/or at the surface of the water. *Phase contrast.*

A   The cells are rounded (spherical or ovoid).   GO TO 59

B   The posterior end of the cell is drawn out into squat arms. Cells 15–50 µm long.

Fig. 112 BRACHIOMONAS

A   Cells with two flagella and one contractile vacuole. Mostly 15–30 µm long.

Fig. 113 CHLAMYDOMONAS

This widely investigated genus is extensively reviewed by Cain (1986) and Harris (1989).

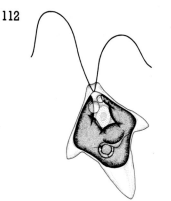

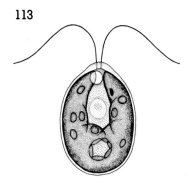

B   Cells with four flagella and two contractile vacuoles. Cell body 7–40 µm long.

Figs 114 & 115 CARTERIA

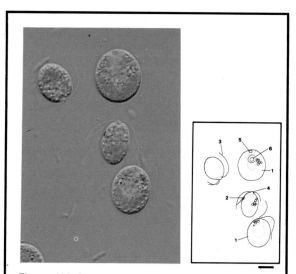

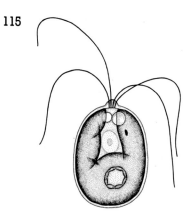

Figure 114 *Carteria.* A coccoid chlorophycean green algae, distinguished from the more familiar *Chlamydomonas* by its four flagella (3), arising together around an anterior protrusion (4). The cells are bright green because of the cup-shaped chloroplast (1). Euglenids have chloroplasts of the same colour, but members of the two groups can be distinguished because chlorophytes have a rigid and unridged cell wall, and because the red eyespot or stigma (2) lies inside the plastid. The plastid encloses the nucleus (6), and the contractile vacuoles (5) discharge anteriorly. *Differential interference contrast.*

A The cell is enclosed by an organic vase, with one long flagellum emerging from a single opening. The lorica may be smooth or have spines, and tends to become brown with age. The loricae of most species are between 10 and 50 μm long.

Figs 116 & 117(a) & (b) TRACHELOMONAS

Trachelomonas is a frequently encountered genus. For taxonomy, see Huber-Pestalozzi (1955), and for structure etc., see Dunlap *et al.*, (1983), Couté and Iltis (1981) and West *et al.* (1980).

116

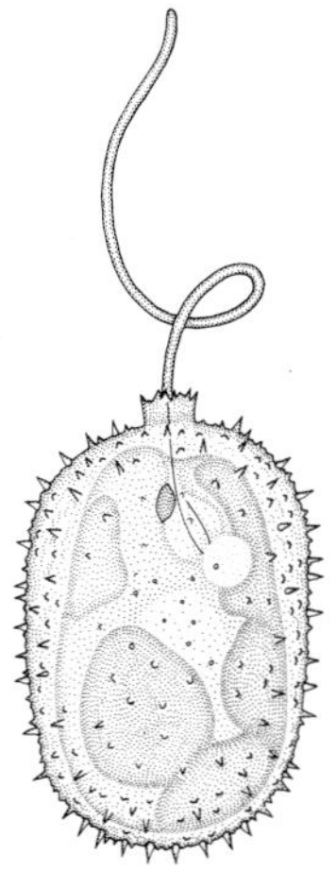

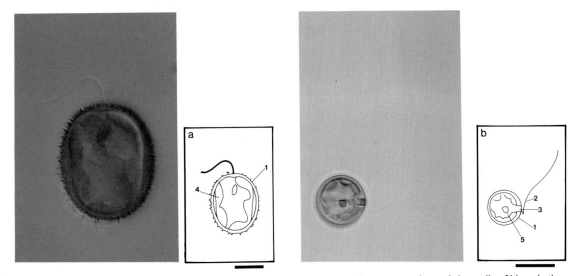

Figures 117(a) & (b) *Trachelomonas*. Two species of this genus of free-swimming, loricated euglenids are shown. They inhabit rigid loricae (1), and a single, long flagellum (2) emerges from an apical aperture (3) in each lorica. The outline of the cell is more evident in the species with the spiny lorica, but in both cases green chloroplasts (4) are evident on the outer surface of the cells. Although the chloroplasts themselves are bright green and contain chlorophyll b, the cells are usually golden or brown owing to the absorption of metal salts by the test. The red organelle (5) is the eyespot (stigma). *Differential interference contrast.*

B The cell is without a lorica.

GO TO 61

A The cell is flexible and is more or less spindle-shaped. Species vary greatly in length, from 20–300 μm.

Euglena is an extensively studied genus, and several books have been dedicated to it alone (Buetow, 1982). The flexibility of the cells is illustrated in Fig. 119, the various shapes being achieved by an active squirming (also called metaboly or euglenoid motion). There is a single emergent flagellum, but this is more evident in Fig. 120, in which the loops that progress along the flagellum are visible. The eyespot or stigma lies outside the chloroplasts. The body surface is spirally sculpted, as seen in Fig. 121. The species illustrated here is without emerging flagella, a state encountered in a number of mud-dwelling species, and in swimming euglenids that have settled against a water–air or water–substrate interface.

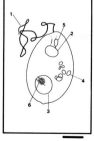

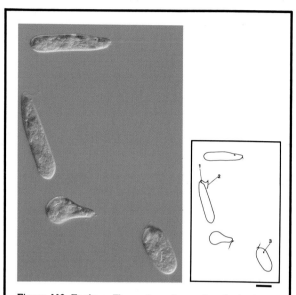

Figure 119 *Euglena.* The various shapes that the body can adopt result from a form of writhing referred to as metaboly or euglenoid motion. This property conveniently distinguishes euglenids with green chloroplasts from other types of green algae. Each cell has a slight anterior indentation (1) where the flagellar pocket opens at the cell surface, and where the single flagellum (2) emerges. The red eyespot (3) can be seen in all of the cells. *Differential interference contrast.*

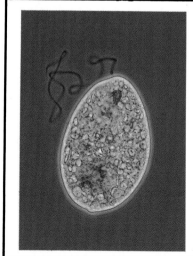

Figure 120 *Euglena.* As with most euglenids, there is one emergent flagellum (1). This is relatively thick, owing to the paraxial rod alongside the axoneme. The body is pliable. Also visible is the region of the flagellar pocket, with an overlying contractile vacuole (2). More posteriorly is the nucleus (3). The surface of the cell is supported by narrow, spiral, interlocking strips, seen in the region over the nucleus (6). The cell is green because it has chloroplasts with chlorophyll b. There are also numerous polysaccharide storage bodies (4), which are refractile and sometimes mistaken for the chloroplasts (cf. Fig. 121). Near the flagellar pocket is the red eyespot (5), involved in sensing the direction and intensity of light. *Phase contrast.*

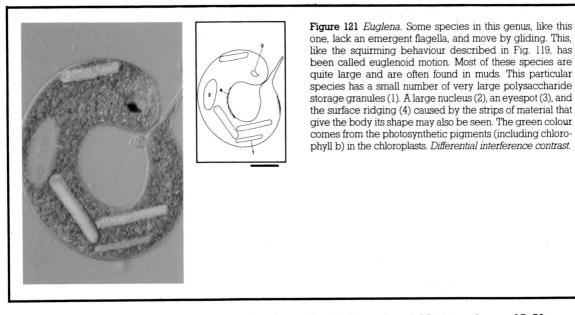

Figure 121 *Euglena*. Some species in this genus, like this one, lack an emergent flagella, and move by gliding. This, like the squirming behaviour described in Fig. 119, has been called euglenoid motion. Most of these species are quite large and are often found in muds. This particular species has a small number of very large polysaccharide storage granules (1). A large nucleus (2), an eyespot (3), and the surface ridging (4) caused by the strips of material that give the body its shape may also be seen. The green colour comes from the photosynthetic pigments (including chlorophyll b) in the chloroplasts. *Differential interference contrast.*

B The cell is not flexible, but is compressed and usually spirally sculpted. Most species are 15–50 μm long. Figs 122 & 123 **PHACUS**

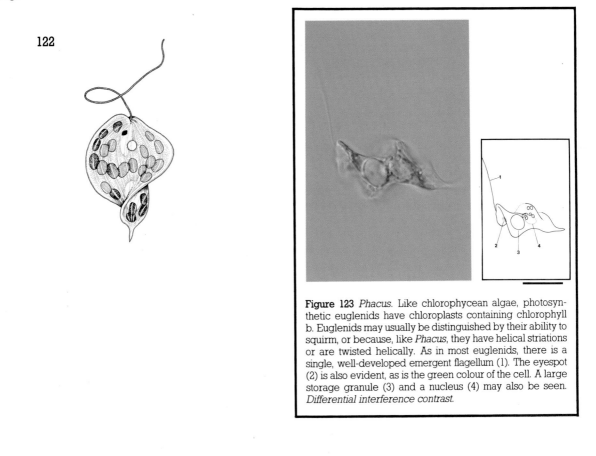

Figure 123 *Phacus*. Like chlorophycean algae, photosynthetic euglenids have chloroplasts containing chlorophyll b. Euglenids may usually be distinguished by their ability to squirm, or because, like *Phacus*, they have helical striations or are twisted helically. As in most euglenids, there is a single, well-developed emergent flagellum (1). The eyespot (2) is also evident, as is the green colour of the cell. A large storage granule (3) and a nucleus (4) may also be seen. *Differential interference contrast.*

62
(54)

A   The cell is red, rounded and rigid, with two flagella of equal length inserting at the apex, and with the cytoplasm running out to the cell wall as fine threads. 20–70 μm long.

Figs 110 & 111 (a) & (b) HAEMATOCOCCUS

See notes to Step 57.

B   Not as 62A.

GO TO 63

63
(62)

A   Brown cells in a smooth-walled or spiky lorica, with a single opening from which a long flagellum emerges. 10–70 μm long.

Figs 116 & 117 (a) & (b) TRACHELOMONAS

See notes after step 60.

B   Not as 63A.

GO TO 64

64
(63)

A   Two flagella of more or less the same length emerge together from a slight depression at the front or anterolateral margin of the cell. Flagella arise in a pocket lined with refractile bodies. 15–80 μm long.

THE PIGMENTED CRYPTOPHYTES   GO TO 65

See notes after Step 48. The cells can swim backwards and this may lead to difficulties in distinguishing the front from the back. Only two genera are keyed out here; for others, see references in the notes after Step 47.

B   Cell with only one flagellum, or, if there are two flagella, they are not of the same length and do not emerge together from a point near the front of the cell.

GO TO 66

65
(64)

A   Small, blue-green cell (about 10 μm) with several chloroplasts.

Fig. 124 CHROOMONAS/CYANOMONAS

B   Yellow-green cell (up to 50 μm).

Figs 125 & 126 CRYPTOMONAS

124

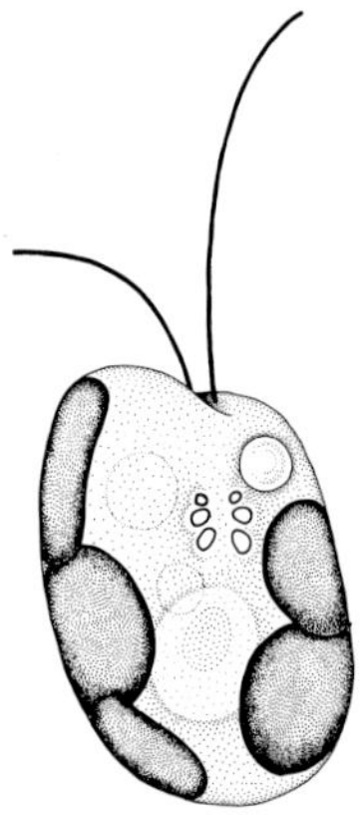

125

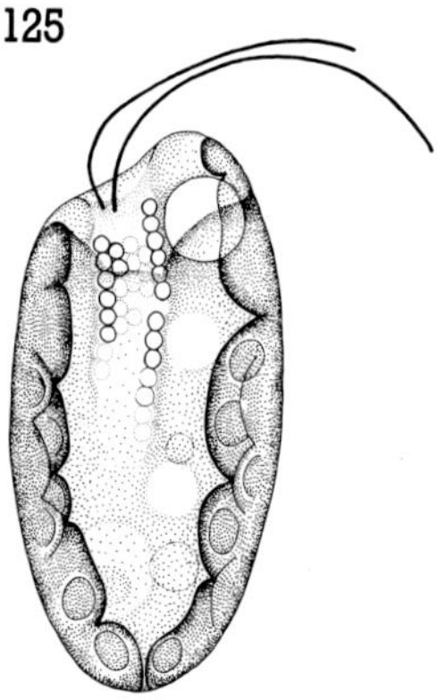

127

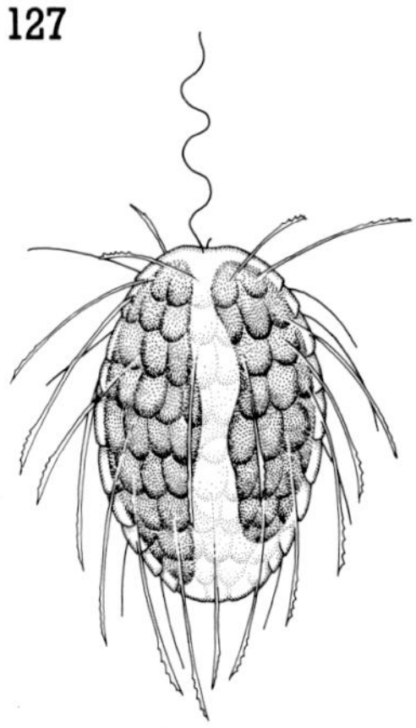

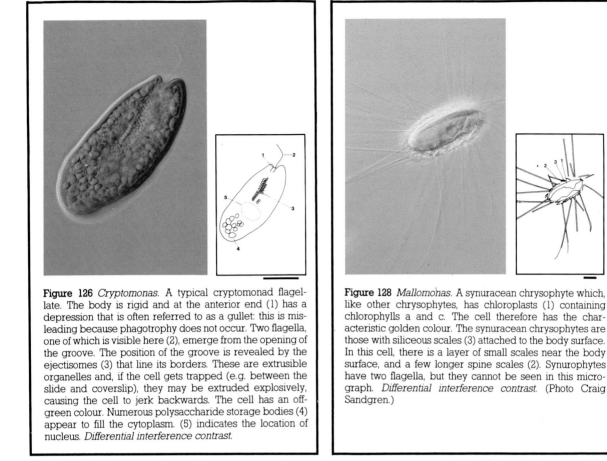

Figure 126 *Cryptomonas*. A typical cryptomonad flagellate. The body is rigid and at the anterior end (1) has a depression that is often referred to as a gullet: this is misleading because phagotrophy does not occur. Two flagella, one of which is visible here (2), emerge from the opening of the groove. The position of the groove is revealed by the ejectisomes (3) that line its borders. These are extrusible organelles and, if the cell gets trapped (e.g. between the slide and coverslip), they may be extruded explosively, causing the cell to jerk backwards. The cell has an off-green colour. Numerous polysaccharide storage bodies (4) appear to fill the cytoplasm. (5) indicates the location of nucleus. *Differential interference contrast.*

Figure 128 *Mallomonas*. A synuracean chrysophyte which, like other chrysophytes, has chloroplasts (1) containing chlorophylls a and c. The cell therefore has the characteristic golden colour. The synuracean chrysophytes are those with siliceous scales (3) attached to the body surface. In this cell, there is a layer of small scales near the body surface, and a few longer spine scales (2). Synurophytes have two flagella, but they cannot be seen in this micrograph. *Differential interference contrast.* (Photo Craig Sandgren.)

A Cell with one flagellum and a layer of scales and/or spines. Species vary between 10 and 70 μm in length.   Figs 127 & 128 MALLOMONAS

66
(64)

Usually regarded as a kind of chrysophyte (see notes after Step 6), scaled forms such as this belong to the Synuraceae. Full identification requires the use of electron-microscopical appearances of scales or spines. In *Mallomonas* only one flagellum is visible; other genera differ in the number of flagella and the character of the siliceous material (Siver, 1991; Starmach, 1985).

B Cell without a coating of scales and spines.   GO TO 67

A The body of the cell is drawn out into several distinct points, and is often large (greater than 50 μm).   Figs 129 & 130 CERATIUM

67
(66)

Ceratium is a planktonic dinoflagellate. Most dinoflagellates have two flagella: one lies inside an equatorial groove (cingulum) that passes around the body, and the other lies in a longitudinal groove (sulcus) and usually trails behind the cell. The cingular flagellum beats with a very shallow amplitude and may be difficult to see. Thus, the cell may seem to have only a single trailing flagellum. One or both flagella are often shed if the cells are illuminated too intensely, or if they are squashed. Mostly planktonic organisms, dinoflagellates tend to be spherical or slightly flattened, and often occur in blooms (natural occurrences of high densities of cells). The nuclei have a peculiar granular consistency due to the arrangement of chromosomes. Cells sometimes have a stigma or eyespot. There are no contractile vacuoles, but some cells have a non-contractile pusule. Some species are phagotrophic (Patterson and Larsen, 1991) and there are some entirely colour-less representatives (Dodge, 1985; Spector, 1985; Taylor, 1986).

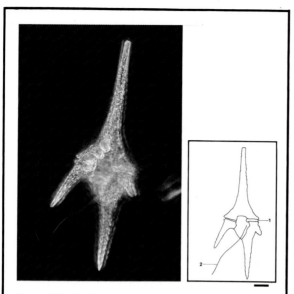

Figure 129 *Ceratium*. A planktonic dinoflagellate, atypically shaped because the body has been pulled out into a number of arms. This is a common genus, encountered in freshwaters and in the sea. The long arms may reduce the amount of energy required to maintain a position within the water column. The equatorial groove (cingulum) is evident (1), as is the trailing flagellum (2). The chloroplasts have the off-green or orange colour that characterizes many dinoflagellates. *Dark ground*.

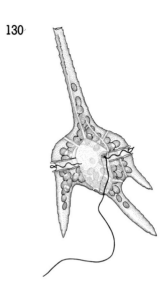

130

B Not as 67A.   GO TO 68

68
(67)

A Cells larger than 10 μm, with a body that is inflexible or has evident stiffening. Large, brown chloroplasts more or less fill the body.   **THE DINOFLAGELLATES   GO TO 69**

See notes after Step 67.

B Small cells (usually less than 10 μm), with one or two golden chloroplasts. Two flagella: one long and extending in front of the cell in a gentle curve, the other short and bending backwards to lie near the cell surface.
                                                                              Fig. 131 OCHROMONAS

The archetypal chrysophyte. See notes after Step 6. For discussion, see Slankis and Gibbs, 1972; Hibberd, 1970).

131

A The cingulum (see notes after Step 67) is near the anterior of the cell (10–30 μm long).

Figs 132 & 133 AMPHIDINIUM

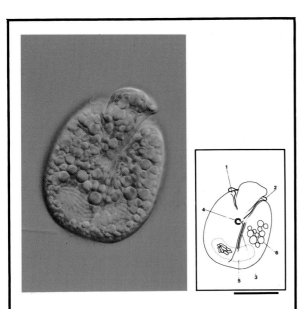

133

Figure 132 *Amphidinium*. A dinoflagellate in which the equatorial groove (cingulum) (1), with its cingular flagellum (2), is near the anterior end of the cell. The cell is consequently split into two unequal parts. Near the junction of the cingulum and the longitudinal groove (sulcus) (3) is a pusule-like organelle (4) of uncertain function. The nucleus (5) lies in the posterior of the cell. The chromosomes in this and other dinoflagellate nuclei are condensed, even when the nucleus is not dividing, and this accounts for the granular appearance of the nucleus. This species has a chloroplast, the colour of which is evident, but the boundaries of which are obscured by refractile cytoplasmic droplets. *Differential interference contrast.*

B The cingulum is near the centre of the cell.

GO TO 70

ALL SCALE BARS 20 μm UNLESS OTHERWISE INDICATED

A The margins of the grooves are well marked by ridges, and the cell surface appears to be divided into plates. Cell length 20–100 µm.

Figs 134 & 135 PERIDINIUM

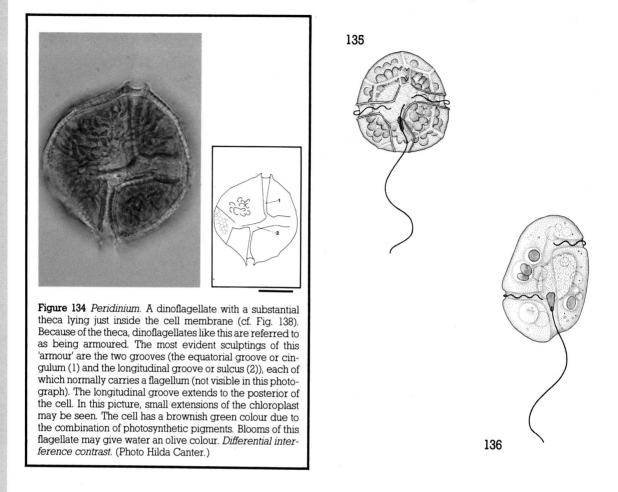

135

136

Figure 134 *Peridinium*. A dinoflagellate with a substantial theca lying just inside the cell membrane (cf. Fig. 138). Because of the theca, dinoflagellates like this are referred to as being armoured. The most evident sculptings of this 'armour' are the two grooves (the equatorial groove or cingulum (1) and the longitudinal groove or sulcus (2)), each of which normally carries a flagellum (not visible in this photograph). The longitudinal groove extends to the posterior of the cell. In this picture, small extensions of the chloroplast may be seen. The cell has a brownish green colour due to the combination of photosynthetic pigments. Blooms of this flagellate may give water an olive colour. *Differential interference contrast*. (Photo Hilda Canter.)

B The test is not divided into plates.

GO TO 71

A The cingulum is oblique, and the two ends do not meet. Most species are 10–50 µm in length.

Fig. 136 GYRODINIUM

B The two ends of the cingulum meet. Cell length usually between 10 and 50 µm.

Figs 137 & 138 GYMNODINIUM

137

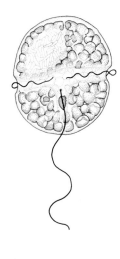

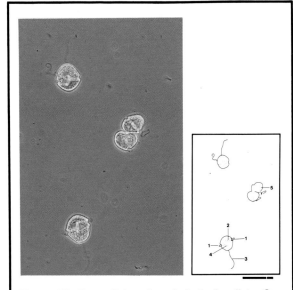

Figure 138 *Gymnodinium*. A typical dinoflagellate. One flagellum (1) usually lies in the equatorial groove (2), but it has become separated in the uppermost cell; a second lies in the longitudinal groove (4), but trails behind swimming cells (3). These cells have an off-green colour because of the combination of pigments in the chloroplasts. A dividing cell (5) is visible. *Differential interference contrast.* (Scale bar 100 μm.)

A Cells without cilia or flagella. Feeding and motion are achieved by cytoplasmic motion and/or by extensions from the cell surface (pseudopodia). AMOEBAE GO TO 73

72 (2)

B Not as 72A (ciliates, suctoria and heliozoa). GO TO 116

Care must be taken to distinguish between the formation of pseudopodia (temporary extensions from the cell); squirming of cells, as may be typical of normal euglenids (which may also shed their flagella, making identification even more difficult); and the distortion that may be encountered in squashed ciliates, which may be very plastic. If you are unsure, try to find more cells to enable you to establish whether the amoeboid form is 'normal'.

The ability to form pseudopodia is widespread among eukaryotes, and the protists that have this ability are not closely related. Major categories of amoebae are distinguished by the shape and number of the pseudopodia, by other morphological features of the cell, such as the uroid, contractile vacuole behaviour, nuclear appearance and nuclear division, and by life cycle phenomena. Amoebae either have many pseudopodia (polypodial) (Fig. 139(a)) or they behave like one pseudopod with a single advancing front (monopodial) (Fig. 139(b)). The pseudopodium may be broad and rounded (lobose) (Figs 139(a) & (b)), usually having a watery leading margin (hyaline cap) (Figs 139(a) & (b), & 210) into which cytoplasmic organelles do not penetrate. Other species may have conical pseudopodia (Fig. 139(d)) or thread-like (filose) pseudopodia (Figs 139(c) & (e)). Some amoebae have a broad advancing front from which fine 'subpseudopodia' project (Fig. 208). Pseudopodia emerge from the anterior and anterolateral margins of moving cells. The posterior end of the cell (uroid) may have diagnostic value, being rounded, lobed or finely folded (compare Figs 196 & 202).

Many amoebae produce loricae or tests. These shells may be organic, with or without adhering material, or they may be formed of secreted inorganic elements. Testate amoebae are usually identified by the appearance of the test. The heliozoa have needle-like pseudopodia, supported internally by stiff axonemes (Figs 139(f) & (g)), and are usually classified as amoebae. The 'axopodia' show little activity, except when observed very carefully. Heliozoa are keyed out elsewhere (Step 191 and the following steps).

General accounts of amoeboid classification are provided by Lee, Bovee and Hutner (1985), Page (1988), and Page and Siemensma (1991).

This key does not deal with very large amoeboid organisms, which are normally referred to as slime moulds (see Lee *et al.*, 1985; Olive, 1975). Many can only be identified when they are in large masses, or after forming spores, but part of their life cycles may involve small amoeboid organisms, which may easily be incorrectly identified as solitary amoebae.

Most smaller amoebae cannot be identified with certainty unless they are isolated and grown in pure culture, so that the different stages of their life cycles can be studied.

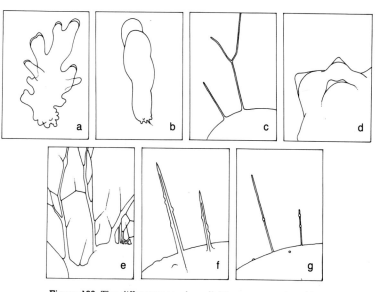

Figure 139 The different pseudopodial types encountered in amoebae: **(a)** polypodial, with many lobose (rounded) pseudopodia emerging from the anterior (top) of the cell. The tips of the pseudopodia have watery hyaline caps, and the posterior end of the cell is crumpled (uroid); **(b)** monopodial cell with a single pseudopodium which has a hyaline cap; **(c)** filose pseudopodia, branching; **(d)** conical pseudopodia; **(e)** filose and anastomosing reticulate pseudopodia; **(f)** actinopod, with central supportive axoneme (as of actinophryid heliozoa); **(g)** actinopod (as of centrohelid heliozoa), with parallel sides and with prominent extrusomes.

73
(70)

A The body is surrounded by an obvious test, and pseudopodia emerge from one or a few apertures.

GO TO 86

B Amoeboid organisms without a rigid test; pseudopodia emerge from many parts of the cell. Body may have adhering scales, other inorganic or organic matter.

GO TO 74

74
(73)

A Amoeboid cells with 'chloroplasts'.

GO TO 75

B Amoeboid cells without 'chloroplasts'.

GO TO 76

Care must be taken to distinguish between chloroplasts, which will always be of the same colour and shape, and ingested algae, which will be of various colours as a consequence of digestion of pigments.

A Amoebae with many bright green 'chloroplasts'. MAYORELLA VIRIDIS

The genus *Mayorella* is illustrated in Figs 87 & 192. The 'chloroplasts' are symbiotic algae. A few species of naked amoebae have similar symbionts (Willumsen, 1982; Page, 1981, 1983), but *M. viridis* is the most common. A colourless species of *Mayorella* is illustrated in Fig. 192.

B Amoebae with a small number of golden chloroplasts. Amoeboid chrysophytes. Cells small, 7–20 μm in diameter.

Fig. 140 CHRYSAMOEBA

See notes after Step 6. Hibberd (1971) gives an account of the fine structure of this organism, and the taxonomy is discussed by Bourrelly (1967).

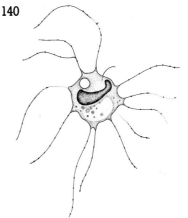

140

A Amoebae with thread-like (filose) hyaline pseudopodia, not emerging from a broad, clearly visible hyaline zone (compare Step 114: pseudopodia are as in Fig. 151, not as in Fig. 208).

FILOSE AMOEBAE   GO TO 77

Some filose amoebae closely resemble heliozoa (Step 187), in that they are spherical and in that the pseudopodia radiate from the body mass. However, the pseudopodia are not rigid (as are heliozoan axopodia), they lack extrusomes (small granules moving along the pseudopodia), and are more transient structures. Fine pseudopodia may or may not fuse (anastomose). Some amoebae, such as *Ac-* *anthamoeba* (see Fig. 208), have fine pseudopodia extending from a broader pseudopodial front. Amoebae with fine pseudopodia are poorly understood and are often assigned to several taxonomic groups, e.g. Filosea, Gromiida, arachnulids, athalamid Granuloreticulosea or the proteomyxids (see e.g. Grassé, 1953; Bovee, 1985b & c; Patterson, 1983, 1984).

B The pseudopodia are not thread-like, but generally broad, although they may have fine sub-pseudopodia arising from a broad hyaline region. GO TO 104

A The filose pseudopodia are visible when the organism is in contact with, and moving over, the substrate. GO TO 78

B The thin pseudopodia are present only in floating cells, but are resorbed after sustained contact with the substrate. Size range great, up to several hundred microns in diameter.

Figs 141 & 142 'AMOEBA RADIOSA'

A. radiosa is not a species of amoeba, rather the floating form adopted by many amoebae. This form is adopted if amoebae are detached from the substrate. The pseudopodia typically taper from a broad base to a narrow tip. The pseudopodia are usually resorbed within a few minutes of settling against the substrate, after which time the normal locomotive form redevelops.

141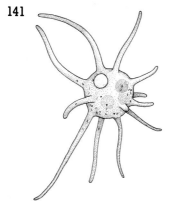

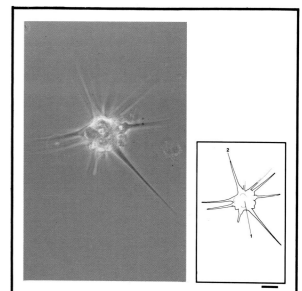

Figure 142 '*Amoeba radiosa*' (or '*Astramoeba radiosa*'). A body shape adopted by many amoebae when they are detached from the substrate. It is a floating form, in which the amoeboid body is contracted to an almost spherical mass (1), from which radiate a small number of tapering pseudopodia (2). Some species may have a single extended pseudopodium, giving them a comet-like shape. For accurate identification, the organism must first be allowed to settle; the diagnostic characteristics will become evident when it starts to move. Confusion with some filose amoebae is also possible. *Phase contrast.*

78
(77)

A Cells with a layer of scales or spicules. GO TO 86

Compare with heliozoa (Step 187) which resemble those amoebae that key out here. Heliozoan pseudopodia have internal skeletal structures (Figs 397 & 406).

B Cells without scales and/or spicules. GO TO 79

79
(78)

A The pseudopodia anastomose (they branch and join up again so that a network is formed). GO TO 80

B The pseudopodia do not branch or, if they do, they do not normally fuse back together again. GO TO 82

80
(79)

A Very large (up to many millimetres) with thin tracts of cytoplasm joining occasional nodes. Often pink.
Fig. 143 RETICULOMYXA

Ostwald (1988) gives a general account of this genus, and other descriptions may be found in Nauss (1949) and Koonce and Schliwa (1985). This is little known taxonomic territory, and identification is very difficult. Soil environments may harbour many large reticulate amoebae, some of which eat fungi.

143

B   Not large (less than 100 μm).   GO TO 81

A   Typically with a single body mass from which very active pseudopodia extend. Many of the pseudopodia appear to be fairly rigid, as if they are internally stiffened. Size range great, up to 600 μm.

Fig. 144 BIOMYXA

B   Many fine, tapering pseudopodia extending from one or more masses or tracts of cytoplasm. The advancing front has a webbed appearance, and the cytoplasm is often granular and orange. Very variable in size: smallest masses of cytoplasm may be 10 μm; cytoplasmic networks may measure many millimetres.

Fig. 145 ARACHNULA

Little is known about these or other naked filose amoebae. *Arachnula* has been described by Dobell (1913) and Old and Darbyshire (1980). These organisms may eat fungi by cutting a hole in the fungal wall (Chakraborty and Old, 1986); they are related to *Vampyrella* (Step 84, Fig. 153), which may attack green algae.

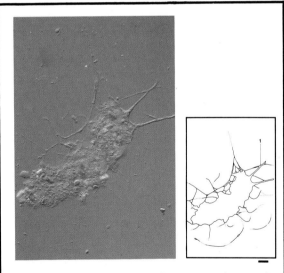

Figure 144 *Biomyxa* (?). With its thin anastomosing pseudopodia (1), this organism falls into a very poorly understood category of amoebae. Organisms with this type of appearance vary enormously in size, but all are distinguished by the intense activity of the cells, with the pseudopodia (1) extending, fusing, separating and resorbing very rapidly. The identification given here is tentative. Affinities are unclear, but there are some similarities with the amoeboid stages of desmothoracid heliozoa (Fig. 413). These organisms are not common. *Differential interference contrast.*

145

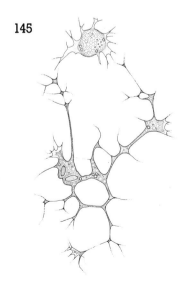

A   Spherical cells.   GO TO 83

B   Cells not spherical.   GO TO 84

A   Very small organisms (less than 10 μm) with one or more orange lipid droplets occupying much of the body.
Figs 146 & 147 DIPLOPHRYS

Fine pseudopodia emerge as two tufts at either pole of the cell, and adhere to the substrate during locomotion. *Diplophrys* is reported to form very large masses, and is covered with a fine layer of scales which can only be seen by electron microscopy. New species have been described recently by Dykstra (1985), and are probably related to the chrysophytes (Green *et al.*, 1989). This obscure genus may have been described under other names, and may have been treated as an alga.

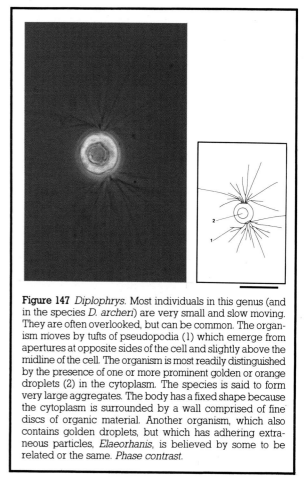

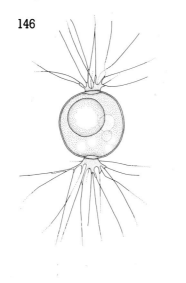

146

Figure 147 *Diplophrys*. Most individuals in this genus (and in the species *D. archeri*) are very small and slow moving. They are often overlooked, but can be common. The organism moves by tufts of pseudopodia (1) which emerge from apertures at opposite sides of the cell and slightly above the midline of the cell. The organism is most readily distinguished by the presence of one or more prominent golden or orange droplets (2) in the cytoplasm. The species is said to form very large aggregates. The body has a fixed shape because the cytoplasm is surrounded by a wall comprised of fine discs of organic material. Another organism, which also contains golden droplets, but which has adhering extraneous particles, *Elaeorhanis*, is believed by some to be related or the same. *Phase contrast.*

B   Size at least 10 μm, often much larger. Colourless except for any food that might have been ingested. Some species have one or more mucus layers around the cell, and pseudopodia emerge from all over the cell surface.
Figs 148–152 NUCLEARIA

There are several recent ultrastructural accounts (Cann, 1986; Cann and Page, 1979; Mignot and Savoie, 1979; Patterson, 1983). The genus includes species that are flattened; some species can be flattened or spherical (Patterson, 1984). Care must be taken to distinguish between flattened forms and amoebae such *Acanthamoeba* (Step 114), in which fine tapering pseudopodia emerge from a broad hyaline zone (subpseudopodia) (Page, 1988). Spherical individuals are easily confused with heliozoa (Step 187).

148

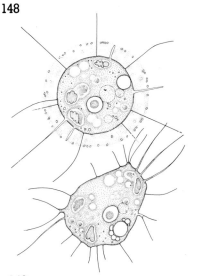

149

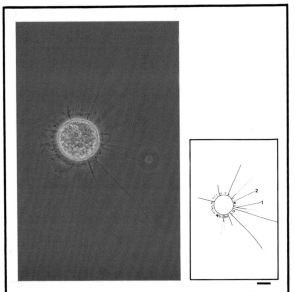

Figure 150 *Nuclearia*. A nucleariid filose amoeba. One of two (unrelated) families normally included in the naked filose amoebae. This species, *N. delicatula*, usually has a rounded body (see Fig. 150) and very long, thread-like pseudopodia (1), so it is easily confused with heliozoa. However, the pseudopodia of the filose amoebae are not stiffened, and there are no extrusomes. The pseudopodia are usually longer at the leading margin of the cell. This species typically has bacteria (2) adhering to a thin layer of mucus. *Phase contrast.*

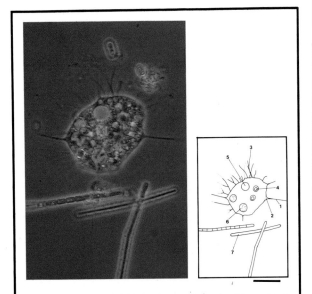

Figure 151 *Nuclearia*. This is a flattened species, *N. moebiusi*, in the same genus that is illustrated in Figs 148 & 150. Some species of *Nuclearia* can adopt both spherical and flattened shapes. In this species, pseudopodia (1) extend from most parts of the cell surface, occasionally arising from an indistinct hyaline region (2). The pseudopodia emanating from the advancing margin of the cell are often long and straight, while those at the posterior end are shorter and folded (3). There is sometimes more than one nucleus (4). Also visible are the contractile vacuole (5), and food vacuoles (6) containing remnants of blue-green algae (7). *Phase contrast.*

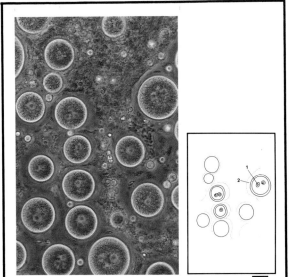

Figure 152 Cysts of *Nuclearia*. These cysts are formed within aggregates of detritus. Each encysted cell may contain one or several nuclei (1) and is surrounded with a thick layer of mucus (2). In this species, cysts form when food is exhausted, but they do not protect the enclosed cell from desiccation (Corliss and Esser, 1974). They presumably serve to extend the survival of the organism when food is absent by decreasing the demands made on food reserves by metabolism. *Phase contrast.*

A   Orange or pink cells, the colour being associated with cytoplasmic granules. Usually associated with green algae, which they penetrate and eat. Cells 10–60 μm in diameter.   Fig. 153 VAMPYRELLA

Vampyrella is probably related to *Arachnula*. It is little studied, but a film is available from the Institut fur Wissenschaftlichen Filmen in Gottingen (Germany) (Hulsmann, 1982).

153

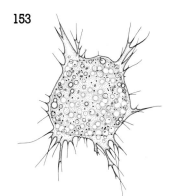

B   Colourless except for any food that might have been ingested.   GO TO 85

A   Small cells (usually less than 10 μm) with fine thread-like, granulated pseudopodia (extrusopodia) emerging from only a few points on the cell surface.   Figs 154 & 155 GYMNOPHRYS

154

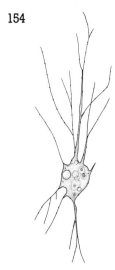

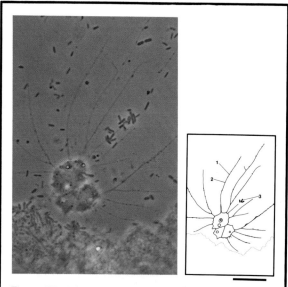

Figure 155 *Gymnophrys*. Small, flattened amoeboid organisms with very long, thin pseudopodia. The pseudopodia (1) may branch and, more rarely, fuse back together again. Very fine granules (extrusomes) (2) may be seen on the pseudopodia. The cells move very slowly. The cell bodies are often found clustered together, embedded in detritus, making it difficult to see individuals. They mostly eat bacteria (3). Flagella have been observed by electron microscopy (unpublished). This organism is rarely reported, but is not uncommon. *Phase contrast.*

B Cells rarely less than 10 μm in length. Pseudopodia emerge from the anterior margin in actively moving cells, or from all margins when stationary. Some species have a layer of mucus over the body, and some also adopt a rounded form.   Figs 148–152 NUCLEARIA

See notes after Step 77.

A Small cells (5–10 μm) with two tufts of fine pseudopodia emerging from opposite ends of the cell. With one or several large orange droplets. The body is covered with a thin smooth organic layer.   Figs 146 & 147 DIPLOPHRYS

See notes after Step 83.

86 (73)

B Body surface not smooth or pseudopodia not emerging as two tufts.   GO TO 90

A Coating apparently formed from agglutinated particles (of sand, etc.), often with lipid droplet in cytoplasm. Body 12–20 μm in diameter.   Fig. 156 ELAEORHANIS

87 (78)

B Coating formed from many similar elements apparently produced by the cell (idiosomes). See Siemensma (1981).   GO TO 88

A Body covered with a loose-fitting layer of long spines with broad, open bases. Very easily confused with heliozoa. Cytoplasm 10–20 μm in diameter.   Fig. 157 BELONOCYSTIS

88 (87)

156

157

B Not as 88A.   GO TO 89

A Body covered with a layer of small siliceous spheres.

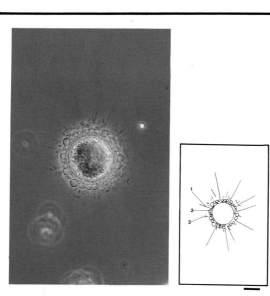

159

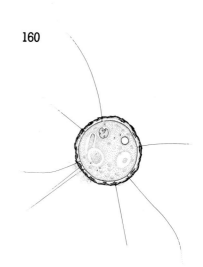

Figure 158 *Pompholyxophrys*. This nucleariid filose amoeba superficially resembles heliozoa, and can be easily confused with them. There are numerous long, thin pseudopodia (1) which lack the stiffness and extrusomes characteristic of heliozoa. The body of the organism is encased within a layer of hollow siliceous spheres (perles) (2). Bacteria may adhere to the outer surface of the perles. The cytoplasm is often orange. Members of the genus feed mostly on algae, and the colour may come from the breakdown of algal photosynthetic pigments. The genus *Pinaciophora* is closely related, but it has flattened plates, not spheres, on the body surface. *Phase contrast*.

B Body covered with a layer of flattened siliceous discs or plates.

Fig. 160 PINACIOPHORA

160

Pompholyxophrys and *Pinaciophora* are often referred to as heliozoa, which they do resemble superficially. Careful observation of the pseudopodia shows that no stiffening elements are present and that extrusomes (small, discrete bodies used for capturing motile prey) as seen in, e.g. Figs 405–409, are absent. They are related to the nucleariid filose amoebae (Patterson, 1985; Page, 1987). Species descriptions are given by Rainer (1969). Recent accounts rely on the electron-microscopical appearance of the siliceous artefacts (e.g. Nicholls, 1983a & b; Nicholls and Dürrschmidt, 1985; Roijackers and Siemensma, 1988; Page and Siemensma, 1991).

A Test around the cell is rigid and has one or more apertures from which the pseudopodia emerge.

GO TO 91

B Test is not rigid, but there is a stiff, flexible sheet (tectum) of fine scales from under which the pseudopodia emerge. With large crystals in the cytoplasm. Body 15–100 μm.

Figs 161 & 162 COCHLIOPODIUM

See Bark (1973) for an account of this genus. Bovee (1985a) refers to seven genera of amoebae with flexible tests, which some specialists regard as testate amoebae. In the strict sense, testate amoebae are those amoebae that have a rigid test (lorica or shell) with usually one aperture (occasionally more) from which the pseudopodia emerge. The testate amoebae form an 'ecological' group which includes organisms that have evolved independently but that look like each other because they occupy similar ecological niches. Testate amoebae are divided into those with lobose pseudopodia and those with filose pseudopodia. General accounts of testate amoebae may be found in Lee *et al.* (1985), Corbet (1973) and Ogden and Hedley (1980).

161

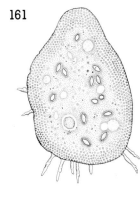

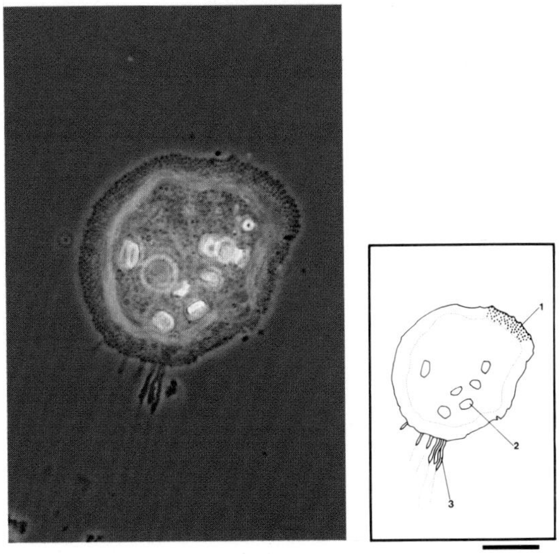

Figure 162 *Cochliopodium*. This amoeba is normally flattened. The dorsal (unattached) surface is covered with fine scales (1), best seen using phase contrast optics. Because of this layer, the genus has often been classified with shelled amoebae. Ultrastructural studies have revealed the presence of a structured organic coating on the outer surface of many other amoebae. The coating is usually too thin to be seen with the light microscope. The 'scales' of *Cochliopodium* may be little more than a particularly well-developed coating. The cell also contains bipyramidal crystals (2), another means by which members of this genus can be recognized. As the cell moves, thread-like strands of cytoplasm may be drawn out behind the cell (3), and short pseudopodia (not visible here) may protrude from under the tectum of scales. Common in freshwater and marine habitats. *Phase contrast.*

A Small cells with a delicate organic test that is usually pressed against the substrate and has several apertures from which thread-like, branching pseudopodia emerge. The pseudopodia bear small particles like those of *Gymnophrys* (Step 85, Figs 154 & 155). Cytoplasmic mass small, usually under 10 μm.

Figs 163 & 164 MICROCOMETES

163

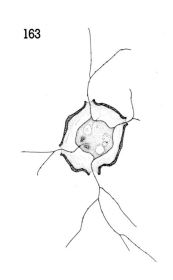

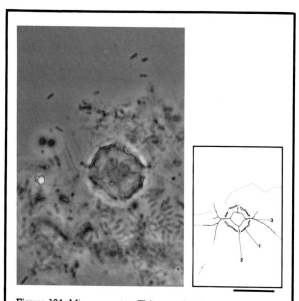

Figure 164 *Microcometes*. This amoeboid organism lives in an organic lorica (1) that becomes brown with age, presumably by absorbing metal salts. The lorica has a number of apertures (2), through which extend very fine thread-like pseudopodia (3). The pseudopodia appear stiff, as if supported internally, and have extrusomes which move along them. *Microcometes* is very similar to *Gymnophrys*, but affinities remain unstudied. The sausage-shaped structures below the amoeba are bacteria. *Phase contrast*.

B   Not as 91A.   GO TO 92

92
(91)

A   Cells with apertures at two poles, from which filose pseudopodia emerge. Cells often with endosymbiotic algae. Test 40–80 μm in length.
                                                       Fig. 165 **AMPHITREMA**

Bonnet (1981a) provides a recent account of the fine structure.

165

B Pseudopodia emerge from a single aperture, located at the apex of the cell or ventrally.   GO TO 93

A Test is smooth and rounded.   GO TO 94

B Test is coloured, textured, or comprised of adhering particles.   GO TO 95

A Test is rigid (it cracks if the organism is squashed), calcareous, and hyaline. The large apical aperture has a slight lip, 10–20 μm. With lobose pseudopodia.   Figs 166 & 167 CRYPTODIFFLUGIA

Bovee (1985a) includes five genera in the family which contains *Cryptodifflugia*. The genera differ in the stength and shape of the test, and in the shape of the aperture. However, Bovee holds that the test is chitinous, in contrast with the more detailed account of Hedley, Ogden and Mordon (1977).

166

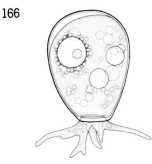

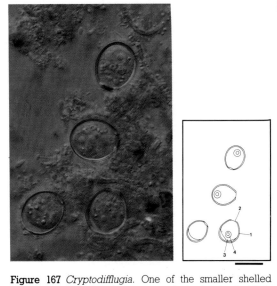

Figure 167 *Cryptodifflugia*. One of the smaller shelled amoebae, which is not uncommon in freshwater habitats and soils. The test (1) is smooth with a single apical aperture (2) which has a slightly incurved, thickened rim. Lobose pseudopodia (not seen here) emerge from the aperture. The cytoplasm is attached to the inside surface of the test by pseudopodia (3). The test is calcareous and brittle, often cracking if crushed by a coverslip. The most prominent organelle is the nucleus (4), with a well-developed nucleolus. This organism is weed-like, sometimes developing in enriched cultures in very large numbers. *Differential interference contrast.*

B Test (20–100 μm long) is organic, and the pseudopodia are fine and branching.
        Figs 168 & 169 **PAMPHAGUS**

The freshwater amoebae with fine pseudopodia and organic tests are very poorly understood. Bovee (1985b) includes over a dozen genera within the filose amoebae. They have been assigned to genera such as *Gromia*, *Chlamydophrys*, *Lecythium*, and *Pseudodifflugia*, but the criteria for assigning organisms to genera are uncertain (Arnold, 1970; Hedley, 1960; Hedley and Wakefield, 1969; Penard, 1902). Similarities also exist with much larger organisms in the genera *Lieberkuhnia* (Fig. 170) and *Allogromia*. These genera have pseudopodia that branch and fuse (anastomose), and thereby resemble marine Foraminifera (Bovee, 1985c). *Allogromia* is usually reported in marine and brackish water sediments.

168

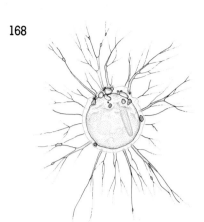

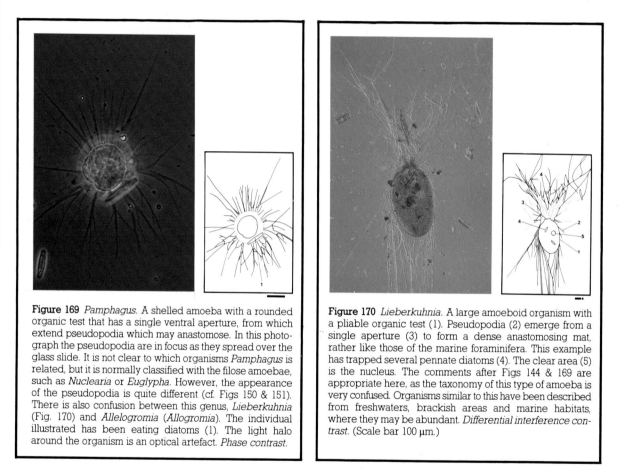

Figure 169 *Pamphagus*. A shelled amoeba with a rounded organic test that has a single ventral aperture, from which extend pseudopodia which may anastomose. In this photograph the pseudopodia are in focus as they spread over the glass slide. It is not clear to which organisms *Pamphagus* is related, but it is normally classified with the filose amoebae, such as *Nuclearia* or *Euglypha*. However, the appearance of the pseudopodia is quite different (cf. Figs 150 & 151). There is also confusion between this genus, *Lieberkuhnia* (Fig. 170) and *Allelogromia* (*Allogromia*). The individual illustrated has been eating diatoms (1). The light halo around the organism is an optical artefact. *Phase contrast*.

Figure 170 *Lieberkuhnia*. A large amoeboid organism with a pliable organic test (1). Pseudopodia (2) emerge from a single aperture (3) to form a dense anastomosing mat, rather like those of the marine foraminifera. This example has trapped several pennate diatoms (4). The clear area (5) is the nucleus. The comments after Figs 144 & 169 are appropriate here, as the taxonomy of this type of amoeba is very confused. Organisms similar to this have been described from freshwaters, brackish areas and marine habitats, where they may be abundant. *Differential interference contrast*. (Scale bar 100 μm.)

95
(93)

A Test does not have adhering particles.   GO TO 96

B Test incorporates inorganic scales or other particles that are all very similar and are therefore probably secreted (idiosomes), and/or includes particles of quartz, diatoms, etc., picked up from the environment (xenosomes).   GO TO 97

A Test is organic, yellow or brown, with a finely reticulate texture, and sometimes with the lateral margin drawn out as a lip with or without dimples or spikes. The aperture is central and ventral. The amoeba has lobose pseudopodia. 40–270 μm.

Figs 171–173 ARCELLA

This genus is common and widespread. For taxonomic reviews, see Decloitre (1976, 1979, 1982). Detailed studies have been made by Netzel (1975a).

171

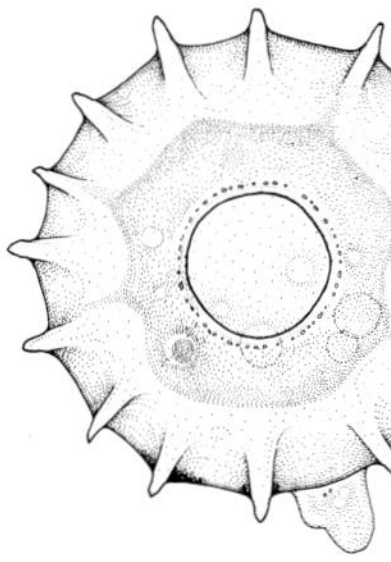

172

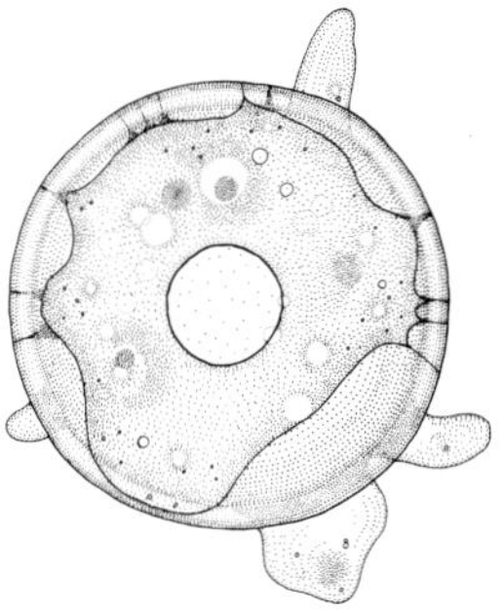

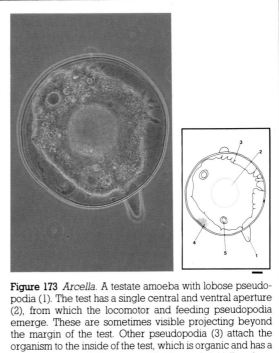

Figure 173 *Arcella*. A testate amoeba with lobose pseudopodia (1). The test has a single central and ventral aperture (2), from which the locomotor and feeding pseudopodia emerge. These are sometimes visible projecting beyond the margin of the test. Other pseudopodia (3) attach the organism to the inside of the test, which is organic and has a very delicate mesh-like texture (4). The test is initially colourless, but it accumulates metal salts from the medium and becomes brown with time. This individual is only slightly impregnated with metal salts. The most prominent organelle in the cytoplasm is the nucleus (5). *Phase contrast*.

B Aperture is located at the end of a slight, tube-like extension of the test, which has a finely dimpled texture. With filose pseudopodia. 30–180 μm.

Fig. 174 CYPHODERIA

The test of *Cyphoderia* is made up of small, adjacent scales, but these cannot easily be distinguished with the light microscope. Bovee (1985a) includes several genera in this group. See also Ogden and Hedley (1980).

174

97 **(95)**	**A** Test is comprised of small, flattish plates.	GO TO 98
	B Material adhering to or comprising the test is not in the form of fine plates.	GO TO 101
98 **(97)**	**A** Aperture is at the end of a slight tubular prolongation of the test, which has very small scales that give it a dimpled appearance. With filose pseudopodia. 30–200 μm long.	Fig. 174 CYPHODERIA
	B Aperture is ventral or apical, and the plates are large.	GO TO 99
99 **(98)**	**A** Component plates are squarish. With lobose pseudopodia. Upt to 150 μm in length.	Figs 175 & 176 QUADRULELLA

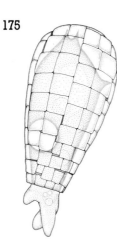

175

Figure 176 *Quadrulella*. A lobose testate amoeba, the test of which incorporates squarish siliceous plates (2). The aperture (1) is apical and, like the test, slightly flattened. A second genus, *Pomoriella*, also with squarish test plates, has a ventral aperture. *Differential interference contrast*. (Photo Helge Thomsen.)

	B Component plates are circular or ovoid.	GO TO 100
100 **(99)**	**A** Aperture is terminal. Some species have spines as well as plates. 30–200 μm.	Figs 177 & 178 EUGLYPHA
	B Aperture is ventral. 20–125 μm.	Figs 179–181 TRINEMA

177

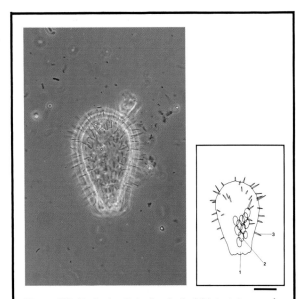

Figure 178 *Euglypha*. Only the shell of this testate amoeba is illustrated. The pseudopodia of living cells are filose and emerge from the single apical aperture (1). The test is comprised of overlapping, flat siliceous scales (2). In this species, some of the scales bear spines. *Euglypha* is one of the more common testate amoebae. *Phase contrast.*

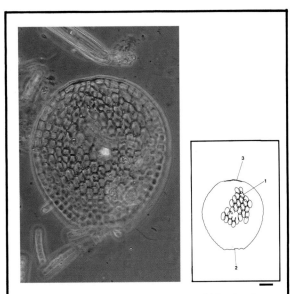

Figure 179 *Assulina*. An empty test, which closely resembles that of *Euglypha* in having numerous overlapping, flattened siliceous scales or plates (1). Unlike *Euglypha*, the test is flattened and the aperture (2) is terminal and slit-shaped. The scales around the aperture (not in focus) have a toothed appearance. Organic matter in the test accumulates metal ions, causing older tests to become brown (3). The amoeba has filose pseudopodia. *Phase contrast.*

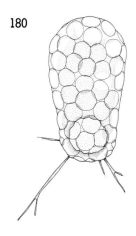
180

Euglypha and *Trinema* have filose pseudopodia. For more detailed studies, see Hedley and Ogden (1974), Hedley *et al.* (1974), Netzel (1972), and Ogden (1979b). Taxonomic reviews are given by Decloitre (1976a, 1979, 1981, 1982). *Assulina* (Fig. 179) resembles *Euglypha*, but there is a substantial organic content to the test, which often looks brown from adsorbed metal ions. *Euglypha* is common in soils and mosses, and plays a significant role in soil ecology.

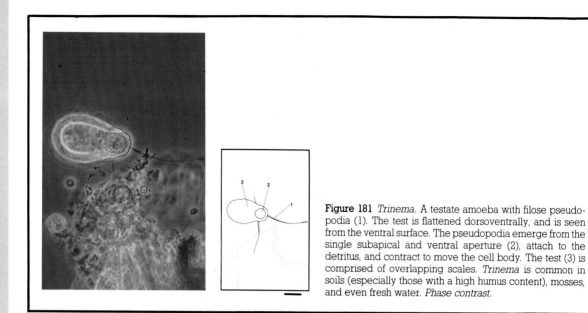

Figure 181 *Trinema*. A testate amoeba with filose pseudopodia (1). The test is flattened dorsoventrally, and is seen from the ventral surface. The pseudopodia emerge from the single subapical and ventral aperture (2), attach to the detritus, and contract to move the cell body. The test (3) is comprised of overlapping scales. *Trinema* is common in soils (especially those with a high humus content), mosses, and even fresh water. *Phase contrast.*

101
(100)

A The test has a spiral shape and incorporates siliceous, sausage-like beads. The mouth is slit-like and ventral. 90–150 μm.

Figs 182 & 183 LECQUEREUSIA

The genus is most usually referred to as *Lesquereusia*. The sausage-like, siliceous beads are not found in all species. The aperture is at the end of a neck, which bends away from the body. Bovee (1985a) includes several genera with this feature, distinguished mostly by the nature and appearance of the adhering material. Harrison *et al.* (1976) give a more detailed account.

182

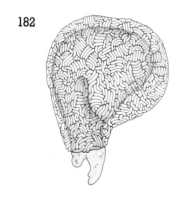

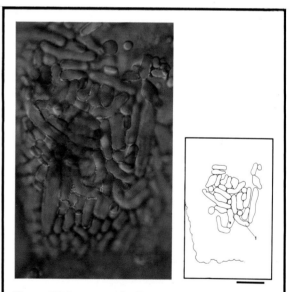

Figure 183 *Lecquereusia*. Testate amoeba. Detail of the test, which incorporates curving siliceous sausages (1) that are bound together and attached to a layer of organic material. The appearance of the siliceous elements is distinctive. The test is slightly flattened and the aperture opens at the end of a short protuberance. This genus is often referred to as *Lesquereusia*. *Differential interference contrast.*

B Not as 101A.

GO TO 102

A Aperture is ventral and near one end of the test; the other end of the test may bear several spines or horns. With irregular adhering particles and lobose pseudopodia. 50–320 μm.

Figs 184 & 185 CENTROPYXIS

For discussions, see Netzel (1975) and Decloitre (1978, 1979).

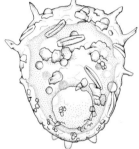

185

Figure 184 *Centropyxis*. A testate amoeba with lobose pseudopodia. Only an empty test is illustrated. The test has a flattened ventral surface, with a single aperture (1) located near the anterior end. The shell is often brown in colour owing to the accumulation of metal ions from the environment by the organic material that forms part of the test. The test may also incorporate small particles picked up from the environment. The posterior end of the test is drawn out into a number of fine spines (2). *Bright field*.

B Aperture is apical.

GO TO 103

A Vase-like test that is round in cross section, and sometimes drawn out as a spike at one end. The test incorporates particles of quartz, etc. picked up from the environment (xenosomes). With lobose pseudopodia. 65–400 μm.

Figs 186–188 DIFFLUGIA

186

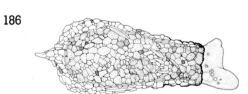

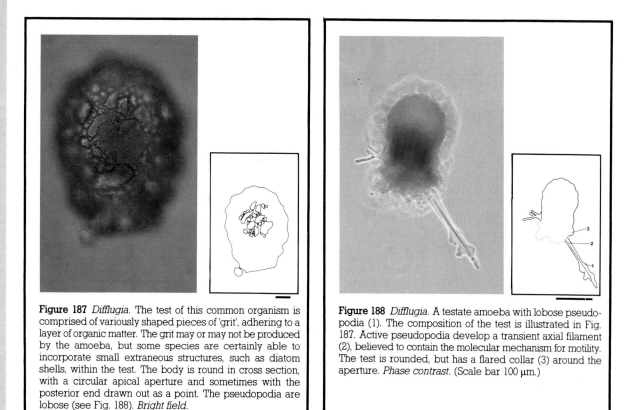

Figure 187 *Difflugia*. The test of this common organism is comprised of variously shaped pieces of 'grit', adhering to a layer of organic matter. The grit may or may not be produced by the amoeba, but some species are certainly able to incorporate small extraneous structures, such as diatom shells, within the test. The body is round in cross section, with a circular apical aperture and sometimes with the posterior end drawn out as a point. The pseudopodia are lobose (see Fig. 188). *Bright field*.

Figure 188 *Difflugia*. A testate amoeba with lobose pseudopodia (1). The composition of the test is illustrated in Fig. 187. Active pseudopodia develop a transient axial filament (2), believed to contain the molecular mechanism for motility. The test is rounded, but has a flared collar (3) around the aperture. *Phase contrast*. (Scale bar 100 µm.)

B Dorsoventrally flattened test, with a terminal aperture. Pseudopodia are lobose, with irregular quartz-grain-like (and/or rounded) particles making up the test. 50–280 µm. Figs 189 & 190 NEBELA

For reviews of these genera, see Decloitre (1977), Netzel (1977), Ogden and Zivkovic (1983), and Bonnet *et al.* (1981b). Motility is discussed by Wohlmann and Allen (1968).

189

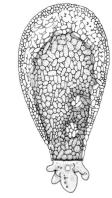

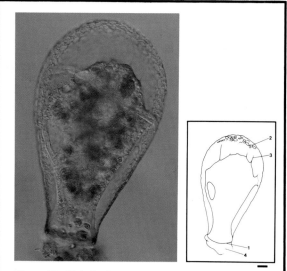

Figure 190 *Nebela*. A testate amoeba. The test is flattened, as is the single apical aperture (1). The margin of the test is structured like a thin flange. The test (2) incorporates siliceous particles produced by the protist. The organism attaches to the inside of the test by means of fine pseudopodia (3). Lobose pseudopodia, or more often an ill-shaped mass of cytoplasm (4), project from the aperture. *Differential interference contrast*.

A The organism is comprised of an anastomosing system of cytoplasmic strands, typically with the anterior aspect broader than the posterior (fan-shaped). From 30 μm to over 1 mm.   Fig. 191 LEPTOMYXA

191

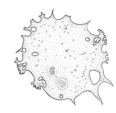 

This multinucleated organism has been described by Pussard and Pons (1976). Uninucleate species with a similar aspect are assigned to the genus *Gephyramoeba* (Pussard and Pons, 1976). For similar marine species, see Page (1983) The reticulate amoebae closely resemble the smaller plasmodial stages of acellular slime moulds (Olive, 1975), from which they may be distinguished because the ebb-and-flow motion of cytoplasm that is characteristic of the true slime moulds is not present. Confident identification requires the isolation of individuals, and their maintenance in pure culture, so that the life cycle may be studied. There is a superficial similarity between *Leptomyxa* and *Arachnula* (see Step 81), and with *Ripidomyxa* (Chakraborty and Pussard, 1985).

B The organism is not comprised of anastomosing strands.   GO TO 105

A When moving, the organism may produce more than one rounded or conical pseudopodium (i.e. is polypodial).   GO TO 106

B The organism moves as if composed of a single pseudopodium, or it has a single pseudopium from which finer cytoplasmic 'subpseudopodia' extend. If there is a single pseudopodium, the cell may move by a sequence of eruptive bulges, forming at the anterior end.   GO TO 108

A The pseudopodia are short, stubby cones, mostly emerging from the broader anterior part of the cell. Common and voracious scavengers. 12–350 μm.   Figs 192 & 193 MAYORELLA

For fuller accounts, see Page (1981, 1983). The plasma membrane is coated with small scales that are visible only with the electron microscope (Pennick, 1975). The organism has a distinctive pattern of contractile vacuole behaviour in which the collapsing vacuole is replaced by a hyaline halo (Patterson, 1981). One species with symbiotic algae (Step 75).

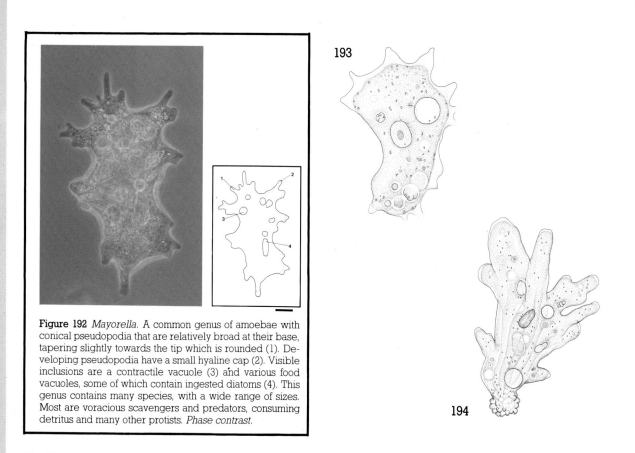

193

194

Figure 192 *Mayorella*. A common genus of amoebae with conical pseudopodia that are relatively broad at their base, tapering slightly towards the tip which is rounded (1). Developing pseudopodia have a small hyaline cap (2). Visible inclusions are a contractile vacuole (3) and various food vacuoles, some of which contain ingested diatoms (4). This genus contains many species, with a wide range of sizes. Most are voracious scavengers and predators, consuming detritus and many other protists. *Phase contrast.*

B   The pseudopodia are hemispherical at their leading edge.

GO TO 107

107
(106)   **A**   Cell with a single large nucleus which has a thick, folded and dimpled wall. Careful observation is needed to see this. Up to 500 μm long.

Figs 194–197 **AMOEBA**

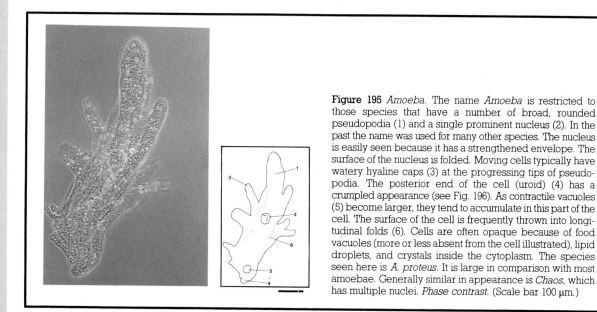

Figure 195 *Amoeba*. The name *Amoeba* is restricted to those species that have a number of broad, rounded pseudopodia (1) and a single prominent nucleus (2). In the past the name was used for many other species. The nucleus is easily seen because it has a strengthened envelope. The surface of the nucleus is folded. Moving cells typically have watery hyaline caps (3) at the progressing tips of pseudopodia. The posterior end of the cell (uroid) (4) has a crumpled appearance (see Fig. 196). As contractile vacuoles (5) become larger, they tend to accumulate in this part of the cell. The surface of the cell is frequently thrown into longitudinal folds (6). Cells are often opaque because of food vacuoles (more or less absent from the cell illustrated), lipid droplets, and crystals inside the cytoplasm. The species seen here is *A. proteus*. It is large in comparison with most amoebae. Generally similar in appearance is *Chaos*, which has multiple nuclei. *Phase contrast.* (Scale bar 100 μm.)

This genus has been extensively studied (Jeon, 1973; Page, 1981, 1984). Many species have been uncritically assigned to this genus, and it has recently been revised to contain only a few species (Baldock *et al.*, 1983; Page, 1981; Page, 1988; Page and Kalinina, 1984).

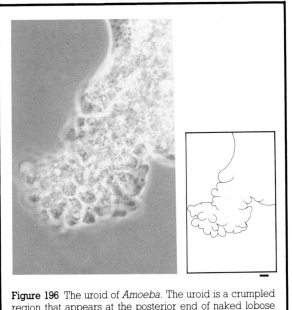

Figure 196 The uroid of *Amoeba*. The uroid is a crumpled region that appears at the posterior end of naked lobose amoebae as they move. It is believed to form as a consequence of the actin–myosin interactions that propel the cell. There are a number of distinctive types of uroid, and they may be used to distinguish different genera and species. This is a morulate uroid. *Phase contrast.*

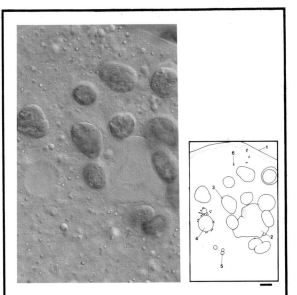

Figure 197 The cytoplasm of *Amoeba*. The plasma membrane (1) lies towards the top of the picture. The most prominent organelle is the nucleus (2), the envelope of which is stiff, has a dimpled texture, and is irregularly folded. This cell has recently ingested a number of *Euglena* cells and a colourless *Polytomella* cell. These are enclosed in food vacuoles (3), around which cluster numerous tiny lysosomes. Lysosomes contain digestive enzymes which are released into the food vacuole as the two organelles fuse. Also visible is a fluid-filled contractile vacuole (4), and around it are smaller vesicles which fuse to fill the vacuole. Mitochondria (slightly larger than lysosomes) cluster around the contractile vacuole. Other visible cytoplasmic organelles are lipid droplets (5) and angular, refractile crystals (6) (a by-product of nitrogen metabolism). All of these organelles are carried around in the flow of cytoplasm as the cell moves. *Differential interference contrast.*

B   Cell with many small nuclei. Up to 1 mm long.

Fig. 198 CHAOS

Superficially very similar to *Amoeba*, this organism is often incorrectly said to have not yet been found in Europe (Siemensma, 1980). *Chaos* is discussed in Jeon (1973) and Page (1976, 1984).

198

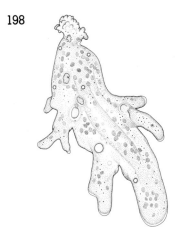

108
(105)

A The organism resembles a single rounded tube (i.e. is monopodial). GO TO 109

B The body is not like a simple tube. Large or small projections may be present, or the wall may be ridged. GO TO 112

109
(108)

A Relatively large (200–2000 μm) and dark cells. The cytoplasm contains refractile quartz grains. The posterior end has a villous bulb uroid. Typically from anaerobic habitats, it moves with cytoplasm flowing forwards along a central axis, and backwards near the margins of the cell. Figs 199 & 200 PELOMYXA

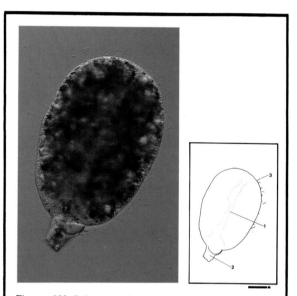

Pelomyxa is now regarded as being related to Mastigamoeba, as it has small, non-motile flagella (Griffin, 1988). Many species have been assigned to this genus, but it probably houses only one (common) species, P. palustris (Margulis et al., 1990; Daniels, 1973), which is said to have a complex life cycle in temperate regions.

200

Figure 199 *Pelomyxa*. A monopodial amoeba. During movement a stream of cytoplasm moves along the central axis of the body (1). A simple uroid (2), which is villous in some individuals, may give rise to some trailing filaments. Short inactive flagella emerge from the cell surface, but these cannot usually be distinguished from bacteria (3) which adhere to the cell. The cytoplasm is usually opaque, as it contains numerous 'sand' grains. Food, mostly filamentous algae, is ingested at the posterior end of the cell. There is probably only one species, *P. palustris*, which varies considerably in size. It is found in anoxic or micro-aerophilic habitats, and is a member of the Pelobiontida. *Differential interference contrast*. (Scale bar 200 μm.)

B Without sand grains. GO TO 110

A Small amoebae (10–35 μm). The hyaline cap is usually easily seen in moving cells. Without a
prominent uroid.   Fig. 201 HARTMANELLA

For more detailed accounts, see Page (1980, 1986).
There may be aggregative phases in the life cycle.

201

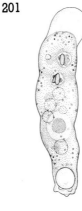

B Hyaline cap usually absent.   GO TO 111

A Small amoebae (usually less than 50 μm) with a villous bulb uroid with thread-like extensions drawn
out to a greater or lesser extent.   Figs 202 & 203 SACCAMOEBA

For further information, see Page and Siemensma
(1991).

203

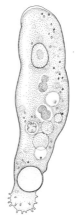

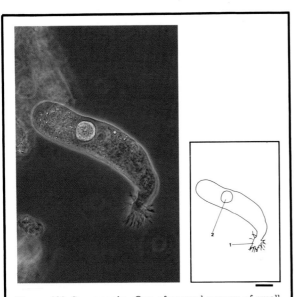

Figure 202 *Saccamoeba*. One of several genera of small
amoebae that have a simple, monopodial appearance.
Hartmannella and *Cashia* are the most likely genera to be
confused with *Saccamoeba*. They may be distinguished by
the appearance of the uroid (1), which in this genus is of the
villous bulb type, and by the presence or absence of the
hyaline cap, which is very much reduced or absent here.
Movement is gradual, i.e. without periodic (eruptive) bulging
of the anterior end of the cell. The inclusion (2) is an ingested
item. *Phase contrast*.

B Small amoebae (mostly 10–20 µm) without villous bulb uroid.

Fig. 204 CASHIA

For further information, see Pussard *et al.* (1980).

204

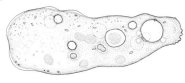

112
(109) **A** Small amoebae (generally less than 30 µm) that are tube-like (monopodial). The pseudopodia develop in a series of eruptive bulges, often to one side of the anterior margin of the cell.

Figs 205 & 206 NAEGLERIA

This diagnosis covers a variety of amoebae (Page, 1988) including the vahlkampfiids, which may produce flagellated stages, and the uninucleate amoebae of the acrasid slime moulds, which have been combined to form the Heterolobosea (Page and Blanton, 1985). Olive (1975) gives an account of the slime moulds. Page (1985) also compares the vahlkampfiid amoebae with the hartmanellids (see Step 110). Two genera of these amoeboflagellates, *Naegleria* and *Tetramitus*, are fairly common and have been keyed out (Step 53) in their flagellate form. One member of the genus *Naegleria* is a facultative pathogen (Martinez, 1985). For more exact identification, pure cultures are required to enable cyst morphology (Pussard and Pons, 1977) or the morphology of the flagellated stage (Page, 1988; Patterson and Larsen, 1991) to be studied.

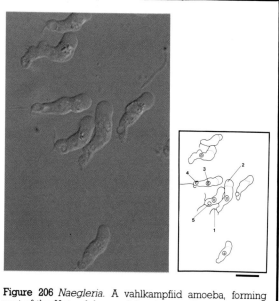

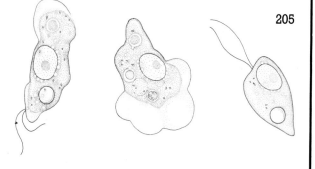

205

Figure 206 *Naegleria*. A vahlkampfiid amoeba, forming part of the Heterolobosea. It has been called an amoeboflagellate, as both amoeboid and flagellated stages may be adopted depending on the prevailing environmental conditions. This photograph shows a number of individuals in the process of resorbing their flagella (1) as they revert from the flagellated to the amoeboid form. Amoebae are monopodial, but they move with the cytoplasm bulging out at the front of the cell in a series of eruptions. Visible components of the cell are the hyaline caps (2), nuclei (3), contractile vacuoles (4) and food vacuoles (5). Some species, especially those from warm water habitats, are facultative pathogens, causing primary amoebic meningitis in humans. They usually enter the body as flagellates, via the nasal passages. Species identification is a specialist procedure and cannot be achieved by light microscopy alone. *Differential interference contrast.*

B Flattened amoebae, with or without fine (sub) pseudopodia emerging from the margins of the cell.

GO TO 113

A Cells with prominent bipyramidal crystals. Some species have an obvious stiff coat (tectum) of fine scales. Cells 15–200 μm long.

Figs 161 & 162 COCHLIOPODIUM

B Not as 113A.

GO TO 114

A Cells with fine pseudopodia extending from a broad hyaline zone. Usually less than 25 μm in size.

Figs 207 & 208 ACANTHAMOEBA

Acanthamoebae are extremely common in soils. Bovee (1985) and Page (1976) describe several genera of amoebae with pseudopodia emerging from a broad hyaline front. These pseudopodia are referred to as 'subpseudopodia' (Page, 1988), and care must be taken to distinguish them from pseudopodia of some filose amoebae, particularly those of *Nuclearia*. One species is mildly pathogenic in man, causing keratitis, an inflammation of the conjunctiva of the eyes, especially under contact lenses. References: de Jonckheere (1983), Martinez (1985), Pussard and Pons (1977).

207

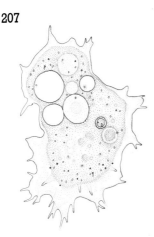

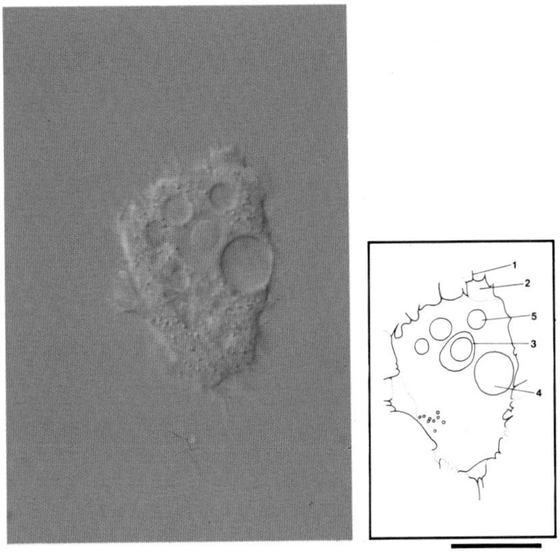

Figure 208 *Acanthamoeba*. An amoeba with filose 'subpseudopodia' (1), which are short, thread-like extensions, distinguishable from the filose pseudopodia of filose amoebae because they arise from a clearly visible hyaline cap (2). Major visible organelles are the nucleus (3), with a prominent nucleolus, and the contractile vacuole (4). Several food vacuoles (5) are present, but they are empty, as this organism was taken from an axenic fluid culture. This genus grows readily under such conditions and has become a popular organism for experimental studies. It is abundant in many soils. At least one species is a facultative pathogen, causing inflammation of the eyes (keratitis), especially in contact lens wearers. *Differential interference contrast*.

B Without subpseudopodia.

GO TO 115

A The dorsal surface of the amoeba (30–200 μm) has slight folds running longitudinally. The hyaline cap is very prominent.

Figs 209 & 210 THECAMOEBA

Systematics are discussed by Page (1976, 1977, 1978) and Singh *et al.* (1981).

209

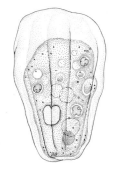

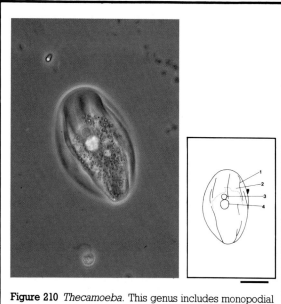

Figure 210 *Thecamoeba.* This genus includes monopodial organisms which are distinguished by the prominent folds (1) that extend more or less longitudinally along moving cells. The hyaline cap region (2) is very prominent. The larger organelles visible inside the cell are the nucleus (3), with its prominent nucleolus, and the contractile vacuole (4). *Phase contrast.*

B Fan-shaped cells (10–80 μm) without folds. The hyaline zone is prominent.

Figs 211 & 212(a) & (b) VANNELLA

References: Bovee, 1965; Page, 1979.

211

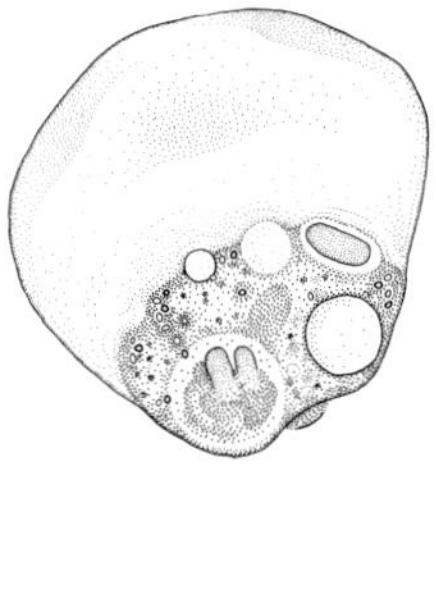

ALL SCALE BARS 20 μm UNLESS OTHERWISE INDICATED

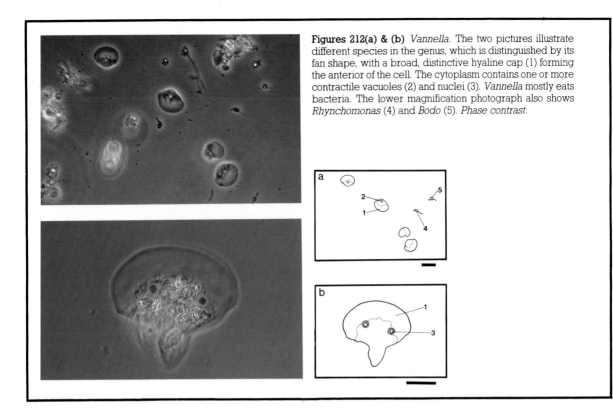

Figures 212(a) & (b) *Vannella*. The two pictures illustrate different species in the genus, which is distinguished by its fan shape, with a broad, distinctive hyaline cap (1) forming the anterior of the cell. The cytoplasm contains one or more contractile vacuoles (2) and nuclei (3). *Vannella* mostly eats bacteria. The lower magnification photograph also shows *Rhynchomonas* (4) and *Bodo* (5). *Phase contrast.*

A Colourless organisms with cilia that are used in locomotion and/or food capture.

THE CILIATES   GO TO 117

116
(72)

The ciliates are a successful group of micro-consumers, and are encountered in most fresh-water habitats. Their large size relative to other pro-tozoa (the majority are 20–200μm long), their fast movements, and the variety of forms ensure that they are often the first (sometimes the only) pro-tozoa to be observed when searching through a sample. Cilia are internally identical to flagella, but they are short in relation to the length of the cell, occur in large numbers, and usually cause fluid to be moved parallel to the cell surface. Cilia may be arranged in clumps, in which individual cilia may be difficult to distinguish. Such aggregates are used for particular types of feeding or motion. It is usual to distinguish between cilia on the general body surface (somatic cilia), and those used in feeding and occurring around the mouth (buccal cilia).

Two of the three major groups of ciliates (the Oli-gohymenophora and Polyhymenophora = spiro-trichs) rely on buccal cilia (as illustrated in Figs 255 & 258) to create currents of water from which sus-pended particles may be isolated and ingested. They are filter feeders, mostly consuming bacteria, but the mechanism has adapted or been superseded in some species to allow the ciliates to feed on larger particles, such as diatoms, other attached particles, or algae. In Polyhymenophora (mostly hypotrichs (Step 136)), heterotrichs (Step 161), and oligotrichs (Step 178) the buccal cilia form a band (the adoral zone of membranelles (AZM)) of stout blocks of ciliary membranelles (Fig. 260). The AZM extends from the anterior of the cell to the site of food inges-tion (cytostome) (Figs 264 & 322).

Oligohymenophora have three membranelles near the cytostome (Fig. 342), but they are usually difficult to distinguish.

A third group, the Kinetofragminophora, is prob-ably polyphyletic and is not well-circumscribed. Many members of this group feed on large part-icles, such as detritus, filamentous algae, animal tissue, or other protists. Predatory species have ex-trusomes associated with the mouth (Fig. 279), and these organelles are used to kill or immobilize prey. Those that consume large algal cells or detritus usu-ally have a tube of stiff rods (nematodesmata: Fig. 379), which is employed to manipulate particles into the cell.

Some ciliates are normally sessile and, of these, some do not have any visible cilia except during 'larval' motile stages (especially suctoria, which are keyed out separately to the ciliated ciliates: see Step 195, Figs 426 & 419).

Somatic cilia lie in rows called kineties (Figs 339 & 359), which run along the length of the body. Loco-motor cilia of crawling species are often confined to the ventral surface, and in hypotrichs (Step 136) they are in the form of aggregates (cirri: Figs 260 & 262).

The evolution of cilia as a behavioural modification of flagella has occurred on several occasions among the protists (e.g. the parasitic opalines, in *Multicilia, Hemimastix*: see Patterson and Larsen, 1991). The true ciliates (Ciliophora) have two kinds of nuclei: the macronucleus and the micronucleus (Figs 348 & 357). In most species, only the former can be seen easily. It may be spherical, ovoid, elongate, or like a string of sausages. Ciliates divide by transverse fission and have a process of sexual reproduction called conjugation (Fig. 344), in which two cells fuse together in the mouth region.

This group is well studied. The biology is reviewed by Corliss (1979), Jones (1974) and Nanney (1980), and evolution by Small and Lynn (1981). The diversity is well covered in Lee *et al.* (1985), Curds (1982), Curds *et al.* (1983) and Kahl (1930–35) with English-language guide by Patterson (1978). Definitive identification of most species and many genera requires special staining of the ciliary apparatus (Fig. 255).

B Ciliates with a strong colour, or cells without obvious cilia (suctoria and heliozoa). GO TO 186

117
(116)

A Cells that swim or crawl around freely; this may include cells normally found in mucus sheaths associated with debris, but which are encountered as free-swimming cells (see *Stichotricha* Figs 218 & 219, *Cyrtolophosis* Fig. 329, *Stentor* Step 118, or *Calyptotricha* Fig. 330). GO TO 131

B Organisms firmly attached to the substrate, or in mucus or other material bonded to substrate, or in a lorica attached to substrate. GO TO 118

118
(117)

A Large (up to 1 mm) cone- or trumpet-shaped cells that can attach at their posterior end. Cilia all over the body, but most prominent at the anterior end. Figs 213–216 STENTOR

Stentors may release their hold and swim away. There may be a tube of mucus around the attached part of the body (Fig. 216). The prominent anterior cilia form an AZM, and somatic cilia cover the body. Many species are coloured (green, blue (Fig. 214) rose, brown, etc.). The large size makes this an amenable experimental organism, and studies on it were reviewed by Tartar (1961). For ultrastructure, see Huang and Pitelka (1973) or Grain (1968), and for an introduction to the taxonomic literature see Warren (1985).

213

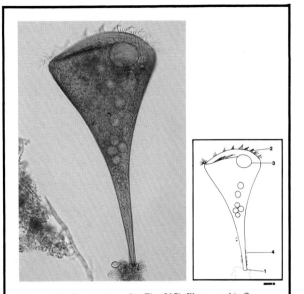

Figure 214 *Stentor* (see also Fig. 215). Illustrated is *S. coeruleus*, the blue species, which is one of the largest in the genus. All species can attach to the substrate by means of a holdfast (1), and, when relaxed and feeding, they have a trumpet shape. Feeding is achieved by the membranelles (2) which run around the flattened anterior end of the cell, i.e. the adoral zone of membranelles (AZM). The contractile vacuole (3) lies adjacent to the cytostome. A few somatic cilia (4) may be seen projecting from the thin part of the cell. *Bright field, closed condenser iris.* (Scale bar 100 μm.)

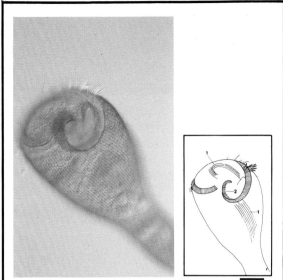

Figure 215 *Stentor* (see also Fig. 214). Details of the arrangement of kineties (1) and of the adoral zone of membranelles of *S. coeruleus*. At its inner end (2), the AZM descends into a buccal cavity. The presence of an AZM indicates that this is a spirotrich (polyhymenophoran), and the somatic kineties reveal it to be a heterotrich. The blue pigment is related to the pink coloration of *Blepharisma* (Fig. 325), and may be physiologically adaptive. *Differential interference contrast.* (Scale bar 50 μm.)

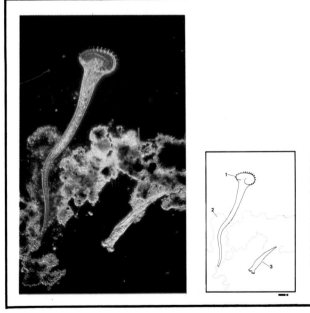

Figure 216 *Stentor*. This is a small, colourless species of *Stentor*. However, like other species, it attaches to the substrate with the narrow posterior end, while the anterior end, with its marginal adoral zone of membranelles (1), becomes disc-like. The beating of the membranelles is co-ordinated, but slightly out of synchrony, resulting in a wave-like profile. This kind of co-ordination is termed 'metachronal'. The basal end of the cell is enclosed within a mucous sheath (2), into which the ciliate may contract. A filter-feeding rotifer (metazoan) is also illustrated (3). *Dark ground.* (Scale bar 100 μm.)

B The body is not evenly covered with cilia.   GO TO 119

A The ciliate occupies a rigid lorica, and may extend a spiral arm that supports the buccal cilia (an AZM). A hypotrich. Body 150–300 µm long.

Fig. 217 CHAETOSPIRA

Other hypotrichs, such as *Stichotricha* (Figs 218 & 219), may be embedded in mucus.

217

218

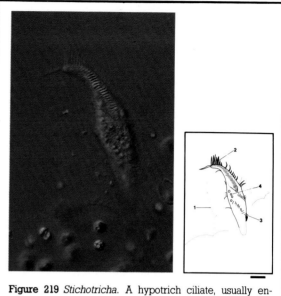

Figure 219 *Stichotricha*. A hypotrich ciliate, usually encountered with its posterior end attached to the debris, and usually in a lorica made of mucus (1). The adoral zone of membranelles (2) leads from the front of the cell to about halfway down the body. The cirri (3) take a spiral path on the body. This individual contains symbiotic algae (4). *Differential interference contrast*.

B Cilia apparently reduced to a band around the anterior end of the body, which is cone-shaped in most species, and is contractile.

THE PERITRICHS   GO TO 120

Peritrichs may occur alone or in colonies. They are common and widespread, being particularly evident in healthy sewage-treatment plants (pp. 190 & 191). For overall accounts, see notes after Step 116, but for taxonomic revisions of particular groups, see Foissner (1979a & b, 1980, 1981), Foissner and Schiffman (1974), Matthes (1981b), Patsch (1974), Roberts *et al.* (1983) and Guhl (1979). The most widely encountered genus, *Vorticella*, is a taxonomic nightmare (see Patsch, 1974, for an account of some more common species). These organisms have evolved from motile Oligohymenophora, and in adapting to a sessile, filter-feeding mode of life, they have lost the somatic cilia. A single posterior band of cilia develops at the base of budding larvae (Fig. 235) to propel them after they separate from the parental cell (Fig. 236), but a similar band develops on trophic cells should the environmental conditions become unacceptable.

Species that attach by a secreted stalk are the sessiline peritrichs. A second group, the mobiline peritrichs (Matthes *et al.*, 1988), have a permanent basal wreath of cilia with which they may move around. The mobiline species are mostly ectosymbionts. To confuse the issue, a few of the so-called sessiline species appear to have secondarily lost the stalk, and they are able to swim around (Figs 373 & 390).

In peritrichs, one of the three membranelles that characterize Oligohymenophora, and the undulating membrane, have been drawn out as the wreath around the anterior of the body (Fig. 234). In profile, these have been likened to two ears (see Leeuwenhoek in Dobell, 1932). One membranelle generates a flow of water, and the extended undulating membrane intercepts particles so that they may be passed to the mouth, which lies at the base of a funnel-shaped cavity. Most sessiline species are contractile, the contractile element forming a coiling strand within the stalk (usually), and extending into the body as a series of fine fibres.

A Species without a lorica, they attach to the substrate by a stalk.　　GO TO 121　　

B Species with a lorica, and/or species without a stalk.　　GO TO 126

A Colonial organisms (more than two cells together).　　GO TO 122　　

B Solitary organisms (may appear as pairs of cells when dividing).　　GO TO 125

At this step, care needs to be taken to distinguish between gregarious peritrichs that retain their individuality but cluster together, and colonial organisms in which the cells are intimately joined to each other. Most peritrich colonies are arborescent, an exception being *Ophrydium*, which forms large gelatinous masses on immersed surfaces (Figs 244 & 245(a) & (b)).

A With contractile stalks.　　GO TO 123　　

B With non-contractile stalks.　　GO TO 124

A The contractile element leading from each cell connects with neighbouring ones. This is most simply determined by jarring the colony so that the peritrichs contract. If the contractile elements are joined together, the entire colony will usually contract simultaneously. Individual cells measure 50–100 µm; colonies may be many millimetres long.　　Figs 220–222 ZOOTHAMNIUM　　

Species in this genus have proved amenable to experimental examination (Suchard and Goode, 1982). General comments are to be found in Wessenberg-Lund (1925), and taxonomy in Bierhof and Roos (1977).

220

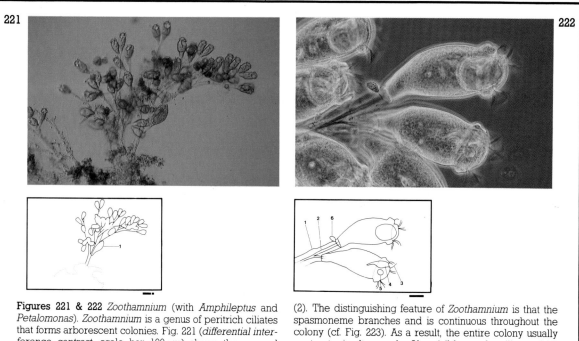

Figures 221 & 222 *Zoothamnium* (with *Amphileptus* and *Petalomonas*). *Zoothamnium* is a genus of peritrich ciliates that forms arborescent colonies. Fig. 221 (*differential interference contrast*, scale bar 100 μm) shows the general appearance of the branching colony. A predatory ciliate, *Amphileptus* (1), moves around the colony, eating individual cells. Fig. 222 (*Phase contrast*). Generic identification of many colonial peritrichs rests upon the details of the stalk (1). Inside the stalk lies the contractile spasmoneme (2). The distinguishing feature of *Zoothamnium* is that the spasmoneme branches and is continuous throughout the colony (cf. Fig. 223). As a result, the entire colony usually contracts simultaneously. Also visible are the cilia used for feeding (3); the buccal cavity (4), into which food is driven before being enclosed in food vacuoles; and a contractile vacuole (5). Crawling across the stalk is a colourless euglenid flagellate, *Petalomonas* (6) (cf. Figs 83 & 84). *Phase contrast*.

B Each cell of the colony can contract independently, the spasmonemes of the different cells not being interconnected. Cells of most species are between 50 and 120 μm long; colonies may extend to several millimetres.

Figs 223–225 CARCHESIUM

See Kahl (1930–35), Foissner and Schiffmann (1974), and general peritrich references for taxonomy.

223

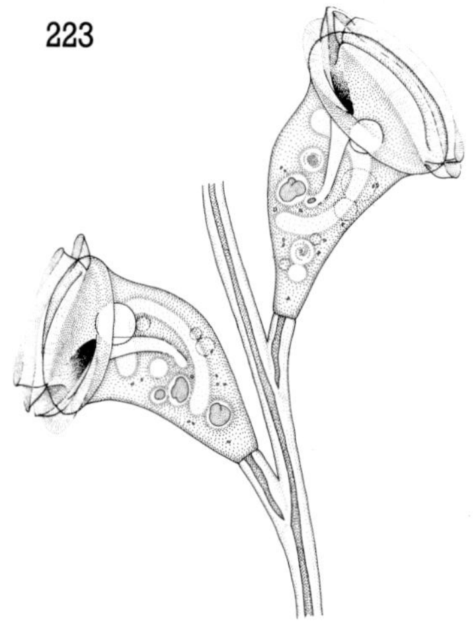

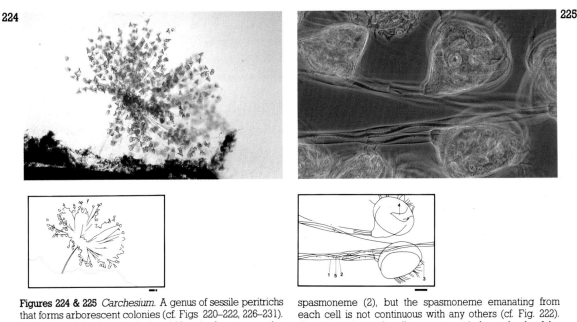

Figures 224 & 225 *Carchesium*. A genus of sessile peritrichs that forms arborescent colonies (cf. Figs 220–222, 226–231). Fig. 224 (*bright field*, scale bar 100 µm), shows an entire colony, with numerous bell-shaped cells, each of which is supported at the end of one branch of the stalk system. In Fig. 225 (*phase contrast*) the stalk (1) has a contractile spasmoneme (2), but the spasmoneme emanating from each cell is not continuous with any others (cf. Fig. 222). Consequently, each cell can contract independently of the entire colony. Also visible are the feeding cilia (1) and the buccal cavity (4).

A The buccal (feeding) cilia extend around a short mushroom-like structure (peristome) which extends from the top of the body. Cell length 25–250 µm. Figs 226 & 227 OPERCULARIA

The taxonomy of this and a similar genus (*Orbopercularia*), which is distinguished by the form of the macronucleus, is discussed by Matthes and Guhl (1977).

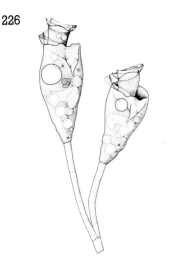

226

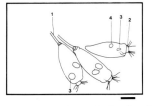

Figure 227 *Opercularia*. A colonial sessile ciliate. The members of the colony are interconnected by a branching stalk system (1). There are no contractile elements within the stalk. These organisms (like a small number of other genera) have their feeding cilia on a small pedestal (2). Also visible are the contractile vacuole (3) and profiles of the macronucleus (4). *Differential interference contrast.*

B There is no protruding peristome. Cell length 50–350 μm, but mostly under 100 μm. Colonies may measure several millimetres.

Figs 228–231 EPISTYLIS

Small and Lynn (1985) include three genera of colonial epistylids.

228

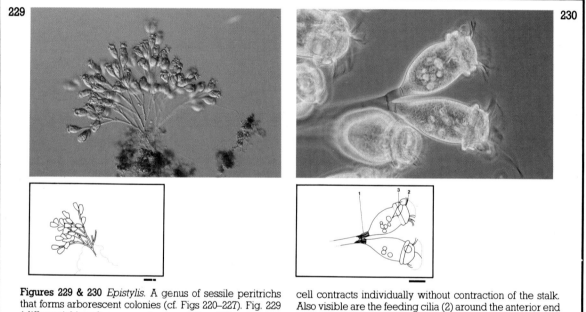

229

230

Figures 229 & 230 *Epistylis*. A genus of sessile peritrichs that forms arborescent colonies (cf. Figs 220–227). Fig. 229 (*differential interference contrast*, scale bar 100 μm) shows an entire colony. Fig. 230 (*phase contrast*) shows a typical striated stalk (1). Members of this genus lack a contractile spasmoneme within the stalk (cf. Figs 222 & 225), and each cell contracts individually without contraction of the stalk. Also visible are the feeding cilia (2) around the anterior end of the cell; the buccal cavity (3), into which food is passed before being enclosed within a food vacuole; and various food vacuoles within the body of the cell.

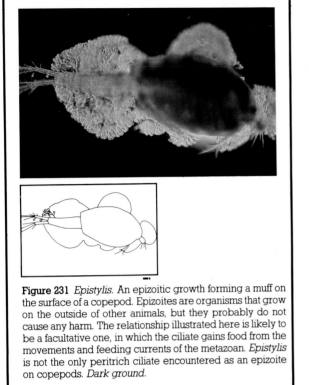

Figure 231 *Epistylis*. An epizoitic growth forming a muff on the surface of a copepod. Epizoites are organisms that grow on the outside of other animals, but they probably do not cause any harm. The relationship illustrated here is likely to be a facultative one, in which the ciliate gains food from the movements and feeding currents of the metazoan. *Epistylis* is not the only peritrich ciliate encountered as an epizoite on copepods. *Dark ground.*

232

A   With a contractile stalk. Length of cell varies from species to species; most are between 20 and 200 µm, most commonly 40–80 µm.

Figs 232–236 VORTICELLA

Vorticella is a very common genus, the taxonomy of which is confused (see notes after Step 119). Some genera other than *Vorticella* will key out here, but special preparative procedures are needed to distinguish them (see Curds *et al.*, 1983)

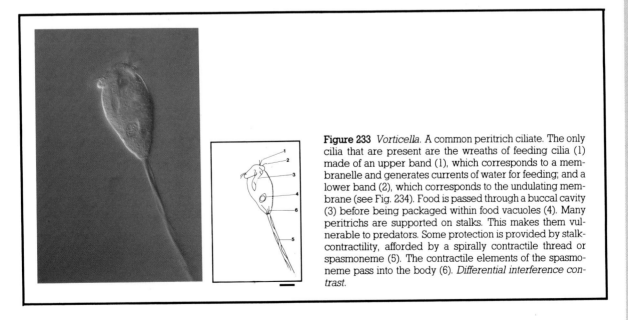

Figure 233 *Vorticella*. A common peritrich ciliate. The only cilia that are present are the wreaths of feeding cilia (1) made of an upper band (1), which corresponds to a membranelle and generates currents of water for feeding; and a lower band (2), which corresponds to the undulating membrane (see Fig. 234). Food is passed through a buccal cavity (3) before being packaged within food vacuoles (4). Many peritrichs are supported on stalks. This makes them vulnerable to predators. Some protection is provided by stalk-contractility, afforded by a spirally contractile thread or spasmoneme (5). The contractile elements of the spasmoneme pass into the body (6). *Differential interference contrast.*

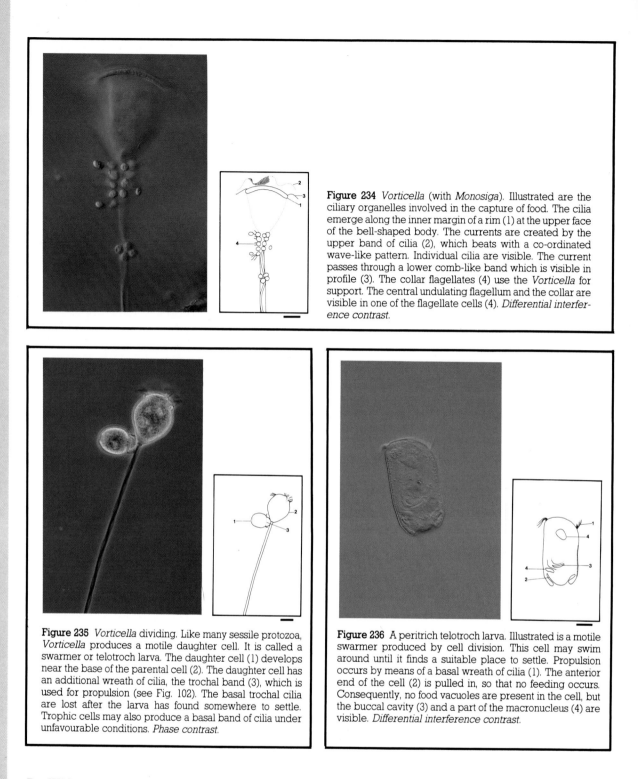

Figure 234 *Vorticella* (with *Monosiga*). Illustrated are the ciliary organelles involved in the capture of food. The cilia emerge along the inner margin of a rim (1) at the upper face of the bell-shaped body. The currents are created by the upper band of cilia (2), which beats with a co-ordinated wave-like pattern. Individual cilia are visible. The current passes through a lower comb-like band which is visible in profile (3). The collar flagellates (4) use the *Vorticella* for support. The central undulating flagellum and the collar are visible in one of the flagellate cells (4). *Differential interference contrast.*

Figure 235 *Vorticella* dividing. Like many sessile protozoa, *Vorticella* produces a motile daughter cell. It is called a swarmer or telotroch larva. The daughter cell (1) develops near the base of the parental cell (2). The daughter cell has an additional wreath of cilia, the trochal band (3), which is used for propulsion (see Fig. 102). The basal trochal cilia are lost after the larva has found somewhere to settle. Trophic cells may also produce a basal band of cilia under unfavourable conditions. *Phase contrast.*

Figure 236 A peritrich telotroch larva. Illustrated is a motile swarmer produced by cell division. This cell may swim around until it finds a suitable place to settle. Propulsion occurs by means of a basal wreath of cilia (1). The anterior end of the cell (2) is pulled in, so that no feeding occurs. Consequently, no food vacuoles are present in the cell, but the buccal cavity (3) and a part of the macronucleus (4) are visible. *Differential interference contrast.*

B With a non-contractile stalk. Cells 20–100 μm long.

Fig. 237 RHABDOSTYLA

Single cells of *Epistylis* (see Step 124) may satisfy this diagnosis. Most reports of *Rhabdostyla* are of ectosymbionts of invertebrates. For taxonomy, see Foissner (1979) or Guhl (1979).

A Without a lorica. The cell attaches directly to the substrate with its narrowed posterior end. There is no stalk or the stalk is indistinct. Cells 20–100 µm long.   Fig. 237 RHABDOSTYLA

237

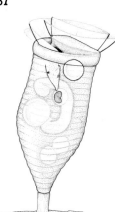

B Cell located in a lorica. Only the anterior portion of the cell extends while feeding.   GO TO 127

A The lorica lies along the substrate, to which it is attached.   GO TO 128

B The lorica is attached to the substrate by its posterior end.   GO TO 129

A Cell attached to a lorica along all or part of its length. Usually with a flap-like structure at the opening of the lorica. Mostly described as ectoparasites from crustacea. Lorica 50–100 µm long.   LAGENOPHRYS

See Kane (1965) for a guide to the genus. Small and Lynn (1985) refer to it as *Circolagenophrys*. See Walker and Roberts (1986) for a recent study.

B Cells attached to the lorica by their posterior ends. Lorica without a flap. Lorica 50–150 µm long.
Fig. 238 PLATYCOLA

See Warren (1982) and Warren and Carey (1983) for guides to the genus.

238

129
(127)

A Lorica with an internal flap which closes behind the organism as it withdraws into the lorica. Cells up to 200 μm in length.

Fig. 239 THURICOLA

See Eperon (1980) for comments on this genus.

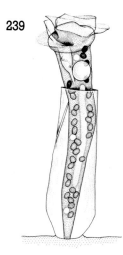

B Lorica without a flap.

GO TO 130

130
(129)

A The lorica has a stalk. Cells of most species are 50–150 μm long.

Figs 240 & 241 COTHURNIA

For a discussion of this genus, see Matthes and Guhl (1973).

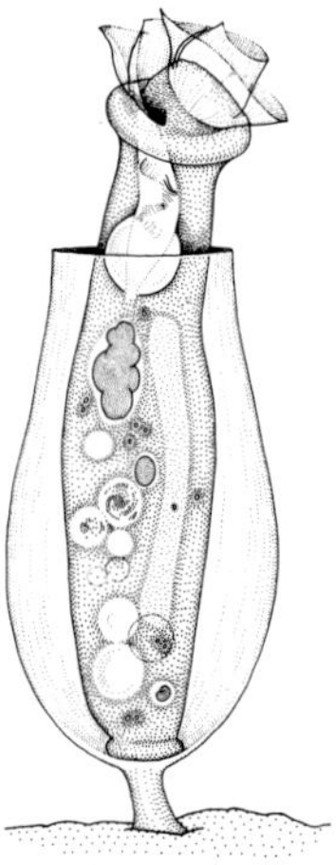

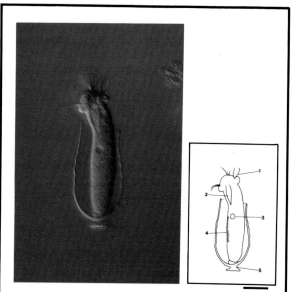

Figure 241 *Cothurnia*. A loricated peritrich ciliate. The only visible cilia are those that form a wreath (1) around the anterior end of the cell. These lead into the buccal cavity (2), where food vacuoles are formed. The contractile vacuole (3) lies alongside the buccal cavity. The surface of the cell is finely striated (4). The lorica is organic and sits on a short foot or pedestal (5). *Differential interference contrast.*

B   The lorica attaches directly to the substrate. Most species are 50–200 μm long.

242

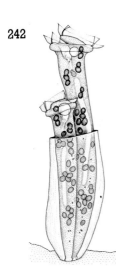

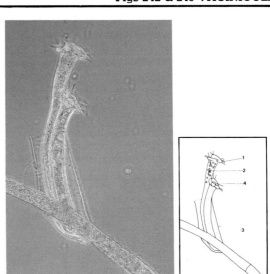

Figure 243 *Vaginicola*. A loricated peritrich ciliate. The only cilia that are evident are those in wreaths at the anterior end. The cytoplasm of this species is green (2) owing to symbiotic green algae. The ciliates project from the lorica (3) when feeding, but can contract into it for protection when stimulated, e.g. by vibration. It is not uncommon for two cells to inhabit the same lorica. A contractile vacuole (4) is visible. The lorica is attached to a filament of the green alga, *Cladophora*. *Phase contrast*.

A   Cells that normally crawl against the substrate.   GO TO 132

Care is needed as many gliding ciliates may swim, although they do not normally do so. It is best to leave preparations for several minutes until cells have settled into their usual patterns of behaviour. In preparations that are drying out or are otherwise compressed, swimming ciliates may be squashed, and thus incorrectly interpreted as gliding. Gliding species are often flattened, presenting one (ventral) surface of their body to the substrate, whereas swimmers may twist or rotate.

B   Cells that do not normally crawl over the substrate, but either swim freely or are embedded in the substrate.   GO TO 154

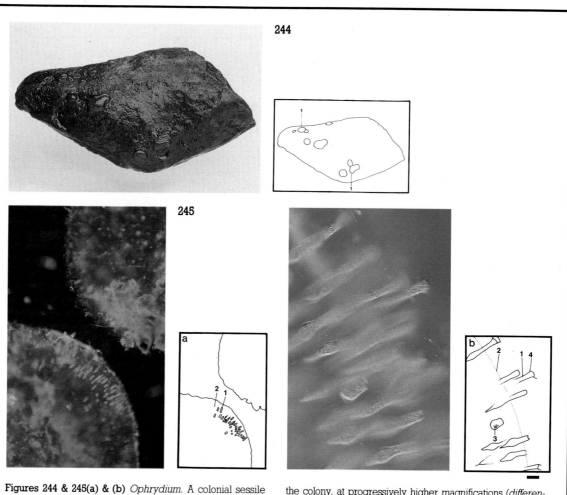

Figures 244 & 245(a) & (b) *Ophrydium.* A colonial sessile peritrich ciliate. The individual cells are embedded in a gelatinous matrix. Fig. 244 shows how the colonies (1) may appear to the naked eye on the surface of a rock (× ½). The two parts of Fig. 245 show the cells (1) within the matrix (2) of the colony, at progressively higher magnifications (*differential interference contrast*). The cells are green because they contain symbiotic green algae (3). The wreath of feeding cilia (4) is visible at the anterior end of some cells.

132
(131)

A Small, flattened cells with a small number of prominent cirri on the ventral surface. Feeding structures are not obvious. Cell length 20–100 μm.

Figs 246 & 247 ASPIDISCA

Aspidisca is a common genus, the dorsal surface of which may be ridged, with the margins pulled out into small spikes. The mouth is comprised of a short patch of membranelles lying in the posterior half of the cell. For taxonomy, see Wu and Curds (1979). A hypotrich (see Step 136).

247

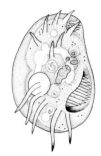

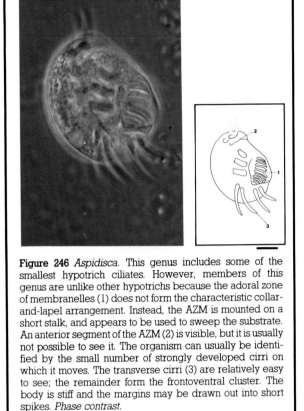

Figure 246 *Aspidisca*. This genus includes some of the smallest hypotrich ciliates. However, members of this genus are unlike other hypotrichs because the adoral zone of membranelles (1) does not form the characteristic collar-and-lapel arrangement. Instead, the AZM is mounted on a short stalk, and appears to be used to sweep the substrate. An anterior segment of the AZM (2) is visible, but it is usually not possible to see it. The organism can usually be identified by the small number of strongly developed cirri on which it moves. The transverse cirri (3) are relatively easy to see; the remainder form the frontoventral cluster. The body is stiff and the margins may be drawn out into short spikes. *Phase contrast*.

B Ciliates that clearly have a mouth, or in which the location of the mouth is easy to identify because of obvious mouth-associated structures. GO TO 133

Mostly with an adoral zone of membranelles (AZM) (mostly hypotrichs), or having a polar or lateral mouth with associated extrusomes for food capture (Kinetofragminophora), or with a ventral mouth surrounded by stout nematodesmata (hypostomes).

A Cells with feeding cilia. These usually lie anterior to, or in, a buccal depression, into which the cytostome opens. GO TO 134

Most species that key out through this step are hypotrichs (see notes after Step 136) or heterotrichs (see notes after Step 161). Both are Polyhymenophora with an adoral zone of membranelles (AZM), evident as a band of parallel lines (the bases of the membranelles) leading from the anterior pole of the cell around the left anterior margin of the cell to terminate at a ventral cytostome (Figs 255, 258 & 261).

B Cells without well-developed feeding cilia around the mouth, but with extrusomes or nematodesmata (rods). GO TO 144

134
(133)

A   Without an adoral zone of membranelles (AZM).   GO TO 135

B   With an AZM.   GO TO 136

135
(134)

A   Flattened cells with a rounded profile, and a large ventral mouth cavity containing a few strongly developed membranelles. Dorsal and ventral body surfaces have 10–20 rows of cilia. Cells between 30 and 150 μm long.
                                                Figs 248 & 249 GLAUCOMA

Fenchel and Small (1980) give an account of the biology of this genus, while Peck (1978) discusses the ultrastructure and Corliss (1971) the taxonomy.

248

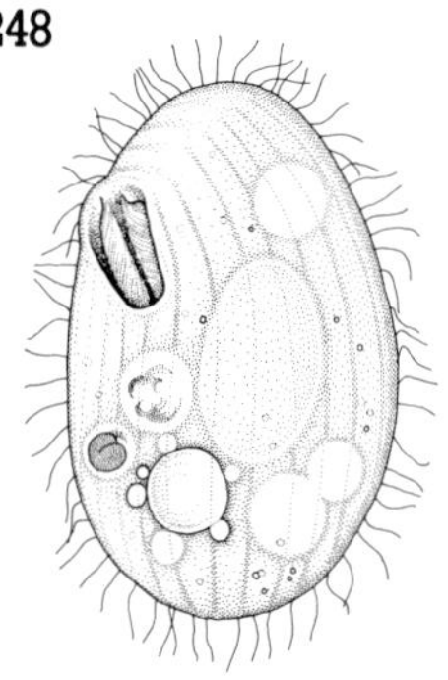

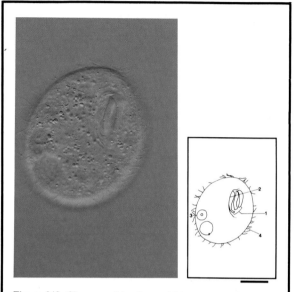

Figure 249 *Glaucoma*. Members of this genus are flattened, with the mouth on the ventral side (as shown here). The three membranelles (1) associated with the mouth are very well developed; they may be used to brush particles of food (which adhere loosely to the substrate) into the mouth. Also visible is the undulating membrane (2), a part of the buccal ciliary apparatus. Near the posterior end of the cell is a contractile vacuole (3), which here has the pore in focus. The organism is evenly ciliated. *Differential interference contrast*.

B   Flattened, rounded cells with a grooved dorsal surface and several long cilia trailing behind. Mouth ciliature not strongly developed. Cells mostly 15–40 μm long.
                        Figs 250 & 251 CINETOCHILUM

Taxonomy is discussed by Jankowski (1968), and structure by de Puytorac *et al.* (1974).

250

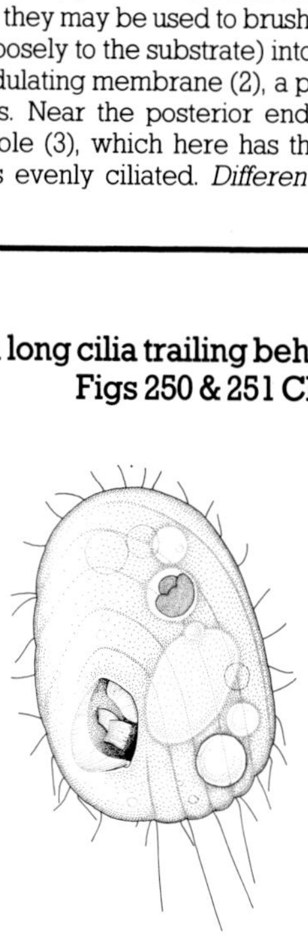

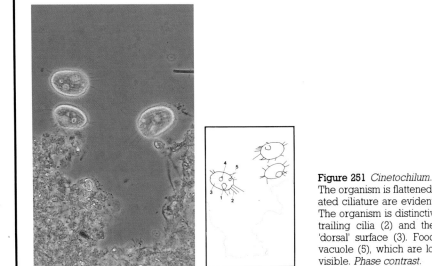

Figure 251 *Cinetochilum*. A common bacterivorous ciliate. The organism is flattened, and the mouth (1) and its associated ciliature are evident in the posterior part of the cell. The organism is distinctive because of the relatively long, trailing cilia (2) and the undulating appearance of the 'dorsal' surface (3). Food vacuoles (4) and a contractile vacuole (5), which are located more posteriorly, are also visible. *Phase contrast.*

A Body with kineties (a heterotrich, see notes after Step 157). May be green with endosymbiotic algae. Cells 100–300 μm long.

136
(134)

Figs 252 & 253 CLIMACOSTOMUM

See Peck *et al.* (1975), Fischer-Defoy and Hausmann (1981) and Foissner (1980a) for accounts of this genus. Other genera in this territory are *Condylostoma* (Fig. 320) and *Peritromus*, which are more usually encountered in brackish or marine sediments.

252

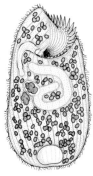

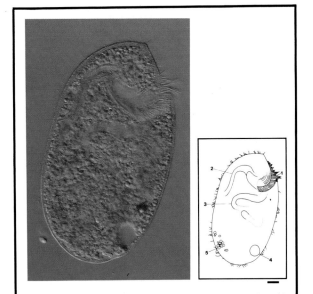

Figure 253 *Climacostomum*. A flattened heterotrich, with the adoral zone of membranelles (1) curving around the front of the cell, and leading into a well-developed buccal cavity (2). A portion of the macronucleus (3) is evident, as is the contractile vacuole (4). The green colour is caused by endosymbiotic green algae in the cytoplasm. However, not all species are green. Despite the presence of algae, this is a voracious feeder, taking in suspended or small particles. *Differential interference contrast.*

B Cells without even body ciliature, but with cirri (blocks of cilia) in groups or rows on the ventral surface of the body.

HYPOTRICHS   GO TO 137

Hypotrichs use the cirri to 'walk' over the substrate. They create currents of water with the adoral zone of membranelles (AZM) and extract bacteria or other small suspended particles, although some species use the AZM to brush diatoms or other items from surfaces, so that they can be eaten. Some species have the cirri arranged in a few small groups (sporadotrichine), but most have them in two or more rows (stichotrichine). Small and Lynn (1985) have segregated these two types, as if to suggest that they are only distantly related.

Hypotrichs are common. There are many genera, and these can only be distinguished by mapping out the location of the cirri. This usually requires silver-staining (Figs 254 & 255). This Guide assumes that silver-staining facilities are not available, and generic identifications must be regarded as imprecise. Different categories of cirri are identified from their location (marginals, frontals, fronto-ventrals, caudals (Fig. 254)). Small bristles may protrude from the dorsal surface. See Borror (1979), Foissner (1982) and Fleury *et al.* (1986) for accounts of the group.

254

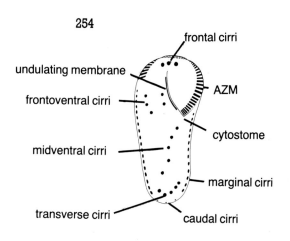

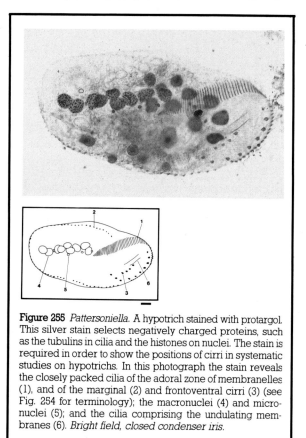

Figure 255 *Pattersoniella.* A hypotrich stained with protargol. This silver stain selects negatively charged proteins, such as the tubulins in cilia and the histones on nuclei. The stain is required in order to show the positions of cirri in systematic studies on hypotrichs. In this photograph the stain reveals the closely packed cilia of the adoral zone of membranelles (1), and of the marginal (2) and frontoventral cirri (3) (see Fig. 254 for terminology); the macronuclei (4) and micronuclei (5); and the cilia comprising the undulating membranes (6). *Bright field, closed condenser iris.*

**137
(136)**

A With stiff bodies.

GO TO 138

B With flexible or contractile bodies.

GO TO 139

A With three prominent tail (caudal) cirri, and marginal rows of cirri that do not form a continuous line around the back of the cell. Body 50–300 μm long.

Figs 256–258 STYLONYCHIA

One of the largest hypotrichs, *Stylonychia* is common. Species are described by Ammermann and Schlegel (1983), and Wirnsberger *et al.* (1985); behavioural biology by Machemer and Deitmer (1987); and electron microscopy by de Puytorac *et al.* (1976).

256

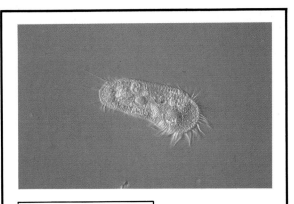

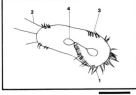

Figure 257 *Stylonychia*. This genus includes some of the largest hypotrich ciliates found in temperate freshwaters. The cell body is rigid, with a well-developed adoral zone of membranelles (1). The genus may often be distinguished by the three long caudal cirri (3). One or two other genera have similar caudal cirri, but in *Stylonychia*, although the marginal cirri (3) form rows along the sides of the body, they do not continue around the posterior end. The two parts of the macronucleus (4) are easily seen. *Differential interference contrast.*

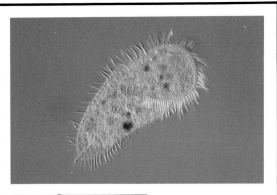

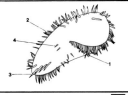

Figure 258 *Stylonychia* (cf. Fig. 257). This view illustrates a number of the features that characterize hypotrich ciliates. They have a well-developed adoral zone of membranelles (1) which forms a collar around the front of the cell, and a lapel leading to the cytostome on the ventral surface. Most species principally have locomotor cilia on the ventral surface, and these are in the form of ciliary aggregates called 'cirri'. In some species, these form rows very much like kineties; in others, as in *Stylonychia*, most or all of the cirri are in clusters. These include marginal cirri (2) at the lateral margins of the cell, an angled line of transverse cirri (3) near the posterior end of the cell, perhaps some caudal cirri (present in this genus, but not illustrated here – see Fig. 257), and patches of frontoventral cirri (4) extending from the anterior part of the body and down the ventral surface. *Differential interference contrast.*

B   Without protruding tail cirri, and without complete marginal rows of cirri. Dorsal surface may be ridged. Body 30–100 μm long.

Figs 259–261 EUPLOTES

Common and widespread, this genus is described by Curds (1975), Gates (1978), and Hill and Reilly (1976). *Uronychia* (Fig. 262), with strongly developed cirri, is related, but is more commonly found in marine or brackish waters.

259

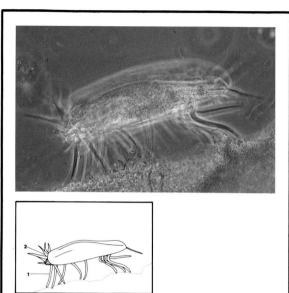

Figure 260 *Euplotes* (cf. Fig. 261). A common and widespread hypotrich ciliate. This unfamiliar view from the side of the ciliate effectively illustrates how the ventral cirri (1) that characterize hypotrichs are used for movement over a substrate. The adoral zone of membranelles (2) draws a current of water under the cell towards the cytostome. Suspended particles may then be removed by the ciliate from the current. *Phase contrast.*

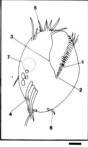

Figure 261 *Euplotes* (cf. Fig. 260). A view of the ventral surface of this hypotrich ciliate. The adoral zone of membranelles (1) forms a collar around the front of the cell, and a lapel leading to the cytostome (2). To the (cell's) right of the mouth is the undulating membrane (3). The locomotor cilia are in several clusters, the most obvious of which is the line of transverse cirri (4). Others include the frontoventrals (5) and the caudal cirri (6). An out-of-focus contractile vacuole (7) is also evident. *Differential interference contrast.*

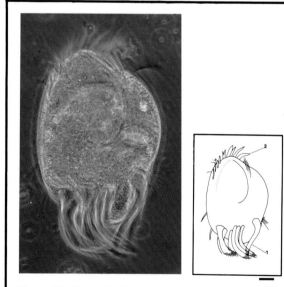

Figure 262 *Uronychia*. One of the more heavily skeletalized hypotrich ciliates. The transverse cirri are extremely well developed (1) and appear to be totally immobile. However, these cells may jump quickly, and the cirri are perhaps brought into use then. This kind of hypotrich (there are several similar genera) does not walk, but swims or glides, presumably using the adoral zone of membranelles (2). *Phase contrast.*

A   Cells with no more than two complete rows of cirri on the ventral surface, although other cirri may be present either singly or in small groups.   GO TO 140

B   More than two complete rows of cirri.   GO TO 141

A   The marginal rows of cirri are continuous around the posterior end of the cell. The other cirri are not in discernible rows. Body 40–200 µm long.   Fig. 263 **OXYTRICHA**

Foissner and Adam (1982) provide an introduction to the diversity in the genus.

263

B   The marginal rows do not meet around the posterior end of the cell. Some species are contractile. Body 50–200 µm long.   Figs 264 & 265 **TACHYSOMA**

264

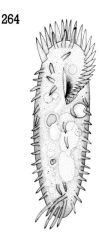

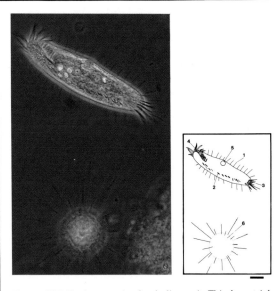

Figure 265 *Tachysoma* (and a heliozoon). This hypotrich has well-developed (immotile) dorsal bristles (1) which might be confused with cirri. Some of the marginal cirri (2) can be seen, and the transverse cirri (3) are very obvious. The adoral zone of membranelles (4) has the typical collar-and-lapel arrangement. A contractile vacuole (5) is visible. Also evident in the picture is a centrohelidian heliozoon (6), recognizable as such by the narrowness of the arms and the prominence of the extrusomes. *Phase contrast.*

141
(139)

A A single line of cirri extends between the marginal rows, from the region at the right of the mouth to more than halfway along the cell. Body 70–300 µm long.   Fig. 266 AMPHISIELLA

See Foissner (1982, 1984) for discussion of this genus.

266

B There is more than one row of cirri in addition to the marginal rows.   GO TO 142

142
(141)

A With two rows (alternating, zigzag pattern) of cirri running along the middle of the ventral surface of the body.   GO TO 143

B Cells with two or more (typically 4–12) ventral rows of cirri in addition to the marginal rows. The macronucleus is in many parts, and the cell has a rounded posterior end. Body 50–400 µm long.
                Fig. 267 UROSTYLA

See Borror (1979) and Foissner (1984c) for a discussion of this genus.

267

143
(142)

A Posterior end rounded. Body 50–200 µm long.   Figs 268–270 HOLOSTICHA

268

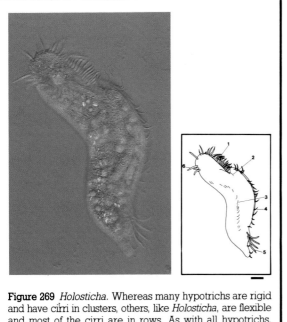

Figure 269 *Holosticha*. Whereas many hypotrichs are rigid and have cirri in clusters, others, like *Holosticha*, are flexible and most of the cirri are in rows. As with all hypotrichs, there is an adoral zone of membranelles (1) in a collar-and-lapel arrangement at the anterior of the cell. The site of the undulating membrane (2) is evident. Most cirri are deployed in three longitudinal rows consisting of one (zigzag) line of midventral cirri (3) and two marginal rows (4). Only one marginal row can be seen clearly here. There is also a cluster of longer, posterior, transverse cirri (5) and a cluster of anterior, frontal cirri (6). *Differential interference contrast.*

Figure 270 *Epiclintes*. A flexible hypotrich commonly found in marine habitats. However, this individual was encountered in sediments from a very low salinity estuarine site. Visible are the adoral zone of membranelles (1), and the marginal (2), midventral (3) and transverse (4) cirri. *Phase contrast.* (Scale bar 100 μm.)

B Posterior end tapering. Body 50–400 μm long.

Figs 271 & 272 UROLEPTUS

There is some debate as to whether ciliates with this form should be placed in one genus or two, the second being *Paruroleptus* (Curds *et al.*, 1983; Borror, 1979).

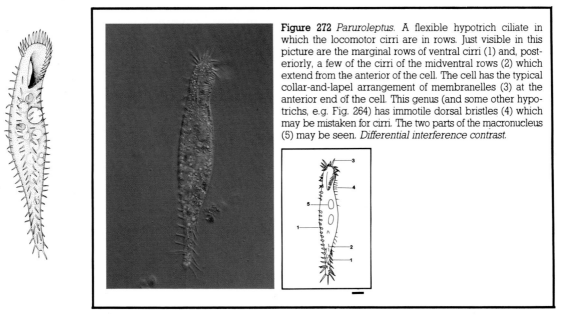

271

Figure 272 *Paruroleptus*. A flexible hypotrich ciliate in which the locomotor cirri are in rows. Just visible in this picture are the marginal rows of ventral cirri (1) and, posteriorly, a few of the cirri of the midventral rows (2) which extend from the anterior of the cell. The cell has the typical collar-and-lapel arrangement of membranelles (3) at the anterior end of the cell. This genus (and some other hypotrichs, e.g. Fig. 264) has immotile dorsal bristles (4) which may be mistaken for cirri. The two parts of the macronucleus (5) may be seen. *Differential interference contrast.*

144
(133)

A Mouth (identifiable by associated cilia, extrusomes or nematodesmata) located at the anterior apex of the cell. GO TO 145

B Mouth located either along the anterior lateral margin of the cell, along the lateral margin, or ventrally. GO TO 147

145
(144)

A The anterior part of the body, which bears the mouth, is rounded in cross section. The body tapers anteriorly, and a wreath of longer cilia may surround the mouth. However, they are not used for sweeping food into the mouth. The rest of the body is evenly ciliated. Body 40–400 μm long.

Fig. 273 TRACHELOPHYLLUM

Trachelophyllum closely resembles *Chaenea* (Fig. 274), the front end of which is stiff.

273

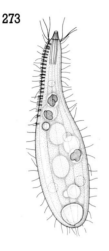

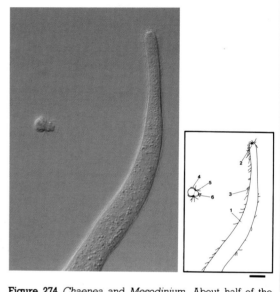

Figure 274 *Chaenea* and *Mesodinium*. About half of the *Chaenea* cell is visible (1). The organism is long and worm-like, with a mouth (2) at the apex of the cell. It is evenly ciliated (3). Alongside is *Mesodinium* (4). This small ciliate is mostly encountered in marine or brackish water habitats, where it is common and widespread, but it is occasionally found in freshwaters. It has a waist region from which two systems of cilia arise: one lies in contact with the posterior half of the cell; the remainder stick out, like cirri. A number of capitate tentacles (6) project from the front of the cell. They may be resorbed if the cell is subjected to mechanical shock. *Differential interference contrast.*

B Mouth region flattened (spatulate). GO TO 146

A The mouth is broadly spatulate, being the widest part of the body. With a single posterior contractile vacuole, and extrusomes underlying the mouth. Body 30–400 µm long. Figs 275 & 276 SPATHIDIUM

Williams *et al.* (1981) and Foissner (1980, 1984) provide an introduction to the appropriate literature.

275

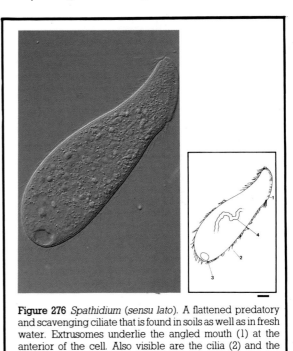

Figure 276 *Spathidium* (*sensu lato*). A flattened predatory and scavenging ciliate that is found in soils as well as in fresh water. Extrusomes underlie the angled mouth (1) at the anterior of the cell. Also visible are the cilia (2) and the contractile vacuole (3). A long, meandering macronucleus (4) is just discernible. *Differential interference contrast.*

B Although the mouth is broadened, it is not noticeably the widest part of the cell. With many contractile vacuoles. Length 150–650 µm. Figs 277–279 HOMALOZOON

The most common species is *H. vermiculare*, studied by Kuhlmann and co-workers (1980) and by Weinreb (1953a & b).

277

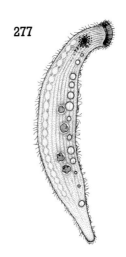

ALL SCALE BARS 20 µm UNLESS
OTHERWISE INDICATED

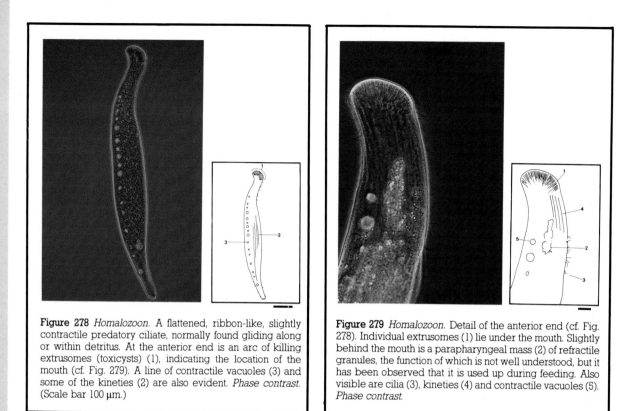

Figure 278 *Homalozoon*. A flattened, ribbon-like, slightly contractile predatory ciliate, normally found gliding along or within detritus. At the anterior end is an arc of killing extrusomes (toxicysts) (1), indicating the location of the mouth (cf. Fig. 279). A line of contractile vacuoles (3) and some of the kineties (2) are also evident. *Phase contrast.* (Scale bar 100 μm.)

Figure 279 *Homalozoon*. Detail of the anterior end (cf. Fig. 278). Individual extrusomes (1) lie under the mouth. Slightly behind the mouth is a parapharyngeal mass (2) of refractile granules, the function of which is not well understood, but it has been observed that it is used up during feeding. Also visible are cilia (3), kineties (4) and contractile vacuoles (5). *Phase contrast.*

147
(144)

A The mouth stretches along an anterolateral margin of the cell. The mouth is usually delineated by the presence of underlying extrusomes.   GO TO 148

B The mouth is located ventrally, and is typically surrounded by a nasse or basket of rods.   GO TO 150

148
(147)

A The mouth takes the form of a concave depression near the front pole of the cell. No extrusomes. Body 100–600 μm long.   Figs 280 & 281 LOXODES

The body is vacuolated, but has no contractile vacuole. Refractile spherules are present (Müller's bodies) and help to orientate the cells (Finlay and Fenchel, 1986). The cell may have a golden colour which is most obvious in the vicinity of the cytopharynx, which leads from the cytostome into the cell. The cells are very flexible, and are from anaerobic habitats, sediments and the water column. For ecology, see Finlay and Fenchel (1986), and for fine structure, see Njiné (1970), and Foissner and Rieder (1983). Often found with *Spirostomum teres* (Fig. 322).

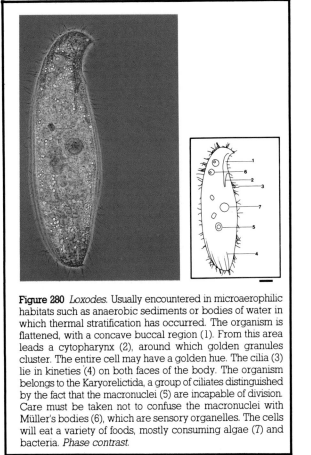

Figure 280 *Loxodes*. Usually encountered in microaerophilic habitats such as anaerobic sediments or bodies of water in which thermal stratification has occurred. The organism is flattened, with a concave buccal region (1). From this area leads a cytopharynx (2), around which golden granules cluster. The entire cell may have a golden hue. The cilia (3) lie in kineties (4) on both faces of the body. The organism belongs to the Karyorelictida, a group of ciliates distinguished by the fact that the macronuclei (5) are incapable of division. Care must be taken not to confuse the macronuclei with Müller's bodies (6), which are sensory organelles. The cells will eat a variety of foods, mostly consuming algae (7) and bacteria. *Phase contrast.*

281

B   The edge that supports the mouth (with extrusomes) is convexly curved.   GO TO 149

GO TO 149

149
(148)

A   The aboral edge of the body bears warts. A long canal extends from the lateral contractile vacuole and along that edge of the body. Macronucleus arranged in a series of interconnected beads. 50–400 μm long.

Figs 282–284 LOXOPHYLLUM

Although quite common, *Loxophyllum* is a little-studied genus. See Fryd-Versavel *et al.* (1976) for a descriptive study, and de Puytorac and Rodrigues de Santa Rosa (1975) for electron microscopy.

282

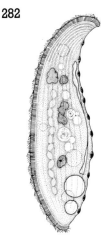

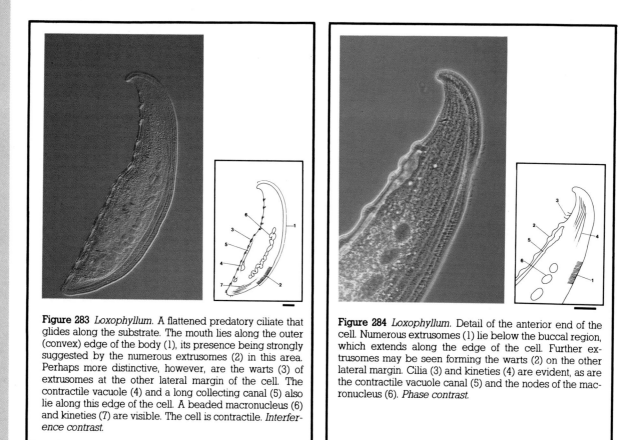

Figure 283 *Loxophyllum*. A flattened predatory ciliate that glides along the substrate. The mouth lies along the outer (convex) edge of the body (1), its presence being strongly suggested by the numerous extrusomes (2) in this area. Perhaps more distinctive, however, are the warts (3) of extrusomes at the other lateral margin of the cell. The contractile vacuole (4) and a long collecting canal (5) also lie along this edge of the cell. A beaded macronucleus (6) and kineties (7) are visible. The cell is contractile. *Interference contrast*.

Figure 284 *Loxophyllum*. Detail of the anterior end of the cell. Numerous extrusomes (1) lie below the buccal region, which extends along the edge of the cell. Further extrusomes may be seen forming the warts (2) on the other lateral margin. Cilia (3) and kineties (4) are evident, as are the contractile vacuole canal (5) and the nodes of the macronucleus (6). *Phase contrast*.

B No warts, with a single posterior contractile vacuole, and usually with the macronucleus in two parts. 40–500 μm.

Figs 285–287 LITONOTUS

Cilia are present in parallel rows on both flat surfaces of the cell. However, in the similarly shaped and behaved *Amphileptus*, the kineties on the upper (right) surface converge to the centre line anteriorly, and those on the ventral (left) surface may be reduced. Another genus, *Hemiophrys*, for a long time erroneously thought to have no cilia on the ventral surface, has been taxonomically merged with *Amphileptus* (Foissner, 1984b). Members of the genus are predatory (Fig. 221). For further references, see Dragesco (1966), Fryd-Versavel *et al.* (1975), and Wilbert and Kahan (1981).

285

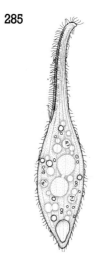

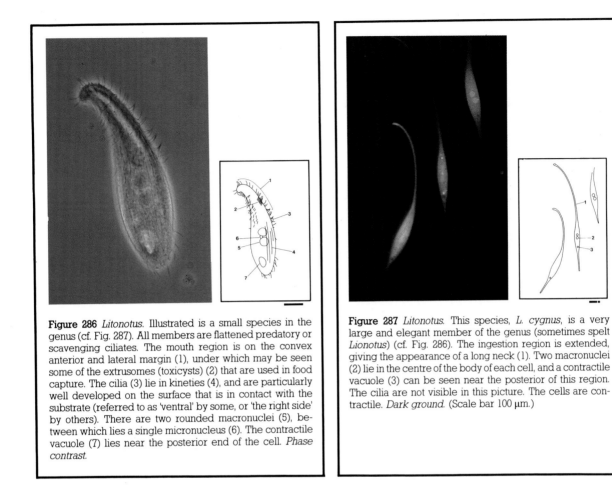

Figure 286 *Litonotus.* Illustrated is a small species in the genus (cf. Fig. 287). All members are flattened predatory or scavenging ciliates. The mouth region is on the convex anterior and lateral margin (1), under which may be seen some of the extrusomes (toxicysts) (2) that are used in food capture. The cilia (3) lie in kineties (4), and are particularly well developed on the surface that is in contact with the substrate (referred to as 'ventral' by some, or 'the right side' by others). There are two rounded macronuclei (5), between which lies a single micronucleus (6). The contractile vacuole (7) lies near the posterior end of the cell. *Phase contrast.*

Figure 287 *Litonotus.* This species, *L. cygnus,* is a very large and elegant member of the genus (sometimes spelt *Lionotus*) (cf. Fig. 286). The ingestion region is extended, giving the appearance of a long neck (1). Two macronuclei (2) lie in the centre of the body of each cell, and a contractile vacuole (3) can be seen near the posterior of this region. The cilia are not visible in this picture. The cells are contractile. *Dark ground.* (Scale bar 100 μm.)

A Cells only slightly flattened. The mouth lies ventrally, and can be seen with care. There are numerous extrusomes and one or two contractile vacuoles, in some cases with distinctive radiating collecting canals. The cells are algivorous, mostly eating diatoms or desmids. Species vary greatly in length, from 50–600 μm.

Figs 288–290 **FRONTONIA**

150
(147)

Mostly found associated with the substrate or detritus, but occasionally encountered in the water column, *Frontonia* is related to *Paramecium*. Both genera have similar star-shaped contractile vacuole complexes. The food may be larger than the ciliate, thus distorting it. The mouth is often difficult to see, as are the rods that surround it. See Didier (1970) for electron microscopy, and Kahl (1930–1935) for descriptions of species.

288

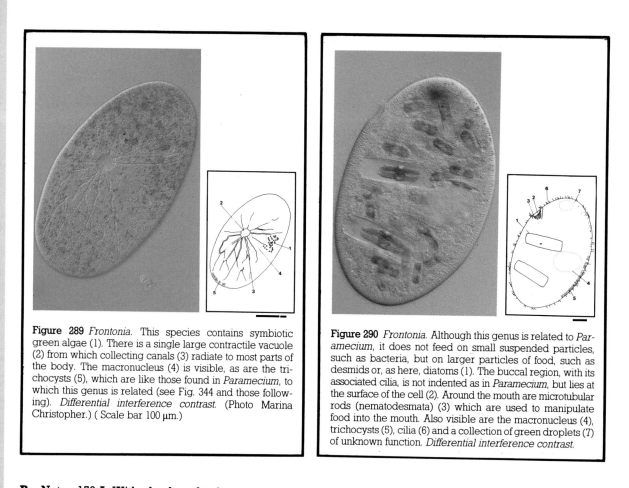

Figure 289 *Frontonia.* This species contains symbiotic green algae (1). There is a single large contractile vacuole (2) from which collecting canals (3) radiate to most parts of the body. The macronucleus (4) is visible, as are the trichocysts (5), which are like those found in *Paramecium*, to which this genus is related (see Fig. 344 and those following). *Differential interference contrast.* (Photo Marina Christopher.) (Scale bar 100 μm.)

Figure 290 *Frontonia.* Although this genus is related to *Paramecium*, it does not feed on small suspended particles, such as bacteria, but on larger particles of food, such as desmids or, as here, diatoms (1). The buccal region, with its associated cilia, is not indented as in *Paramecium*, but lies at the surface of the cell (2). Around the mouth are microtubular rods (nematodesmata) (3) which are used to manipulate food into the mouth. Also visible are the macronucleus (4), trichocysts (5), cilia (6) and a collection of green droplets (7) of unknown function. *Differential interference contrast.*

B Not as 150 A. With a basket of rods around the mouth, and/or cells distinctly flattened. GO TO 151

151
(150)

A The body is flat, with a ridged surface. Extrusomes are associated with each ridge. There is one central contractile vacuole (with vesicular ampullae which appear shortly before the vacuole collapses). The mouth is located near one edge of the cell. Feeds on blue-green algae, and in well-fed cells the mouth may be difficult to see. 50–100 μm long.

Figs 291 & 292 PSEUDOMICROTHORAX

P. dubius has been studied in depth by Hausmann and Peck (1978), and Peck (1985).

291

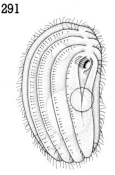

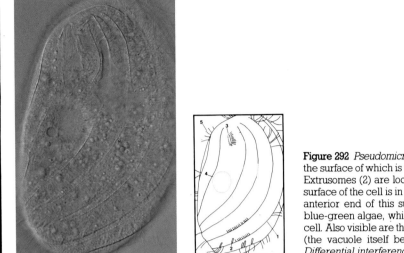

Figure 292 *Pseudomicrothorax*. A flattened algivorous ciliate, the surface of which is thrown into well-developed folds (1). Extrusomes (2) are located along the grooves. The ventral surface of the cell is in focus, and the mouth (3) lies near the anterior end of this surface. The cell ingests filamentous blue-green algae, which accounts for the colour inside the cell. Also visible are the region of the contractile vacuole (4) (the vacuole itself being out of focus) and the cilia (5). *Differential interference contrast.*

B No evident ridging or extrusomes. GO TO 152

A With a cross-banded, ridged groove parallel to and near the margin of the cell. Cilia limited to ventral surface. Some species with an orange 'eyespot'. Mostly eating diatoms. 80–150 µm long.

Figs 293 & 294 **CHLAMYDODON**

For a discussion of this genus, see Kaneda (1960).

293

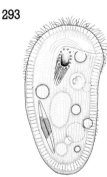

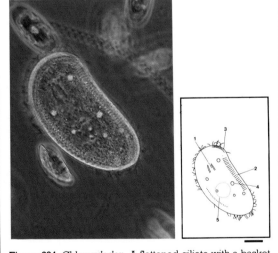

Figure 294 *Chlamydodon*. A flattened ciliate with a basket or nasse (1) of microtubular rods (nematodesmata) opening on the ventral surface of the cell. Members of the genus use the nasse to manipulate larger particles of food, such as diatoms, into the cell. The genus is distinguished by the striated band (2) that lies near the margins of the ventral surface, and by an anterior orange spot (3). Contractile vacuoles (4) and the region of the macronucleus (5) are evident. *Phase contrast.*

B Without a striated groove.

GO TO 153

A With a posterior adhesive spike. Small, mostly under 50 μm in length.

Figs 295 & 296 TROCHILIA

295

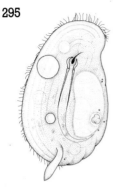

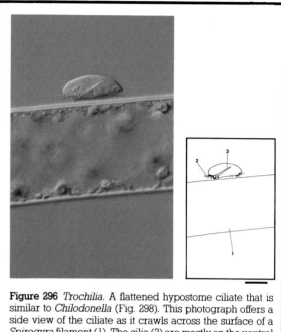

Figure 296 *Trochilia*. A flattened hypostome ciliate that is similar to *Chilodonella* (Fig. 298). This photograph offers a side view of the ciliate as it crawls across the surface of a *Spirogyra* filament (1). The cilia (2) are mostly on the ventral surface. A basket or nasse (3) opens on the ventral surface of the cell, where it is best placed to detach adhering bacteria and other particles of food. *Differential interference contrast.*

B Without a posterior adhesive spike. Mostly 40–120 μm long.

Figs 297 & 298 CHILODONELLA

The previous three genera are typical members of the hypostomes; mostly they consume bacteria or small attached algae. The biology and systematics of the group have been described by Deroux (1970, 1976a, b & c). *Chilodonella* has an unciliated mid-ventral patch, while the closely related *Trithigmostoma* (Fig. 299) does not (Jankowski, 1967b; Foissner, 1988; Hofmann and Bardele, 1987).

297

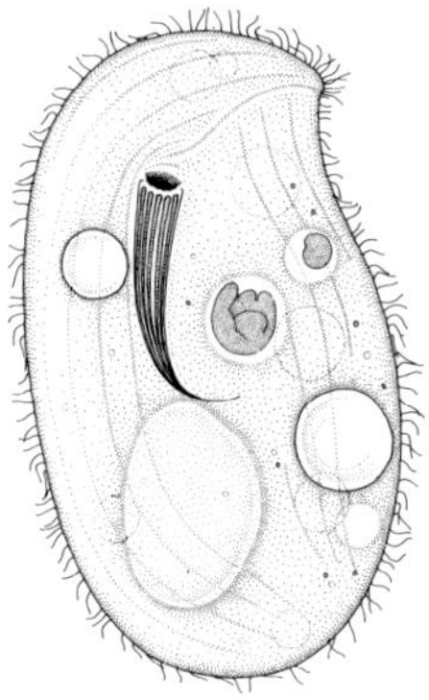

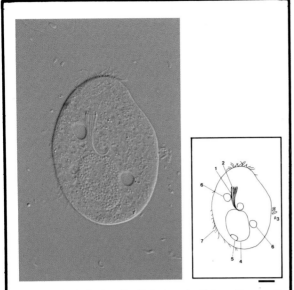

Figure 298 *Chilodonella*. A flattened ciliate with a well-developed basket or nasse (1) of microtubular rods (nematodesmata). At their anterior ends, the rods have 'teeth' (2) which surround the cytostome on the ventral side of the cell (see Fig. 299). For this reason, the group of ciliates to which *Chilodonella* belongs has been referred to as the hypostomes. The rods are used to manipulate bacteria (3) or other small particles into the mouth. Also evident in this picture are the macronucleus (4) with an adpressed micronucleus (5), two contractile vacuoles (6) and the cilia (7). *Differential interference contrast.*

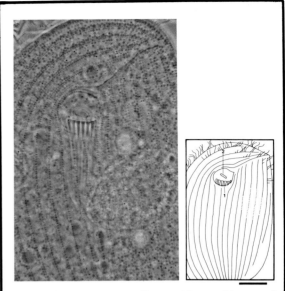

Figure 299 *Trithigmostoma*. Illustrated is part of the ventral surface of this hypostome ciliate. It is closely related to *Chilodonella* (Fig. 298), but as the photograph is a surface view, only the anterior tips of the microtubular nematodesmata (1) around the cytostome (2) are visible. The picture shows the bases of the individual cilia and the distribution of kineties on the ventral surface, with slightly larger dark mitochondria lying near some of the anterior kineties. The contractile vacuoles and macronucleus are not in focus. *Differential interference contrast.*

A Spindle-shaped cells with a long, highly mobile neck supporting a rounded mouth at the end, usually embedded in detritus. The cell may extend up to 1 mm. Figs 300–302 LACRYMARIA

154 (123)

In one common species, *L. olor*, the extensible neck appears to probe for food. Smaller species with less extensible necks are usually placed in the genus *Phialina* (Fig. 303). See Tatchell (1981) and Bohatier *et al.* (1970) for structure, and Foissner (1983) for taxonomy.

300

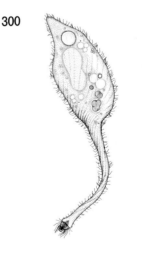

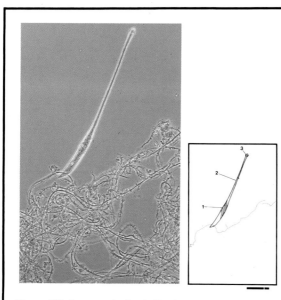

Figure 301 *Lacrymaria*. A spindle-shaped cell, shown here with the 'neck' (2) characteristically greatly extended and with the 'body' drawn out (2) (cf. Fig. 302, which is of the same organism). The 'mouth' is located in the expanded region at the tip of the neck (3). *Lacrymaria* extends and withdraws the neck as it probes for food (here mostly iron bacteria), usually with the posterior end of the body attached to the detritus. *Phase contrast.*

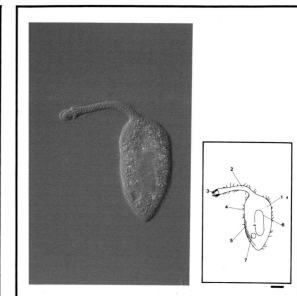

Figure 302 *Lacrymaria*. This organism is normally found with the expanded 'body' (1) of the cell embedded in detritus, and with the anterior 'neck' (2) darting in and out. The neck, which is remarkably extensible, probes the water and the substrate for suitable food. The organism feeds on other protozoa or detritus. At the tip of the neck is the 'mouth' (3) armed with extrusomes for catching prey. The body is ciliated (4), with the cilia lying in spiralling kineties (5). A large macronucleus (6) lies in the centre of the cell, and a contractile vacuole (7) is visible at the posterior end. *Differential interference contrast.* (Scale bar 100 μm.)

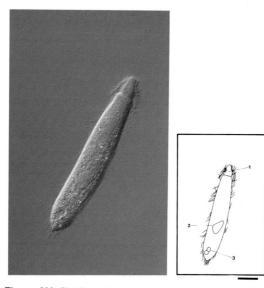

Figure 303 *Phialina*. A torpedo-shaped ciliate, round in cross section, and in many ways very similar to *Lacrymaria* (Figs 301 & 302), from which it has only recently been distinguished. There is an anterior differentiated mouth region (1) with extrusomes and a cluster of long cilia. Inside the body, a macronucleus (2) and contractile vacuole (3) are visible. Unlike *Lacrymaria*, the cell does not have a contractile neck. *Differential interference contrast.*

ALL SCALE BARS 20 μm UNLESS OTHERWISE INDICATED

B   Not as 154 A. Cells usually moving freely through the water. GO TO 155

A   Rounded cells, moving in bounds separated by short pauses. No rows of somatic cilia, but with a row of equatorial spikes. An adoral zone of membranelles (AZM) surrounds the anterior end of the cell. 20–50 μm long. **155 (154)**

Figs 304 & 305 **HALTERIA**

See Grain (1972), Tamar (1968, 1974) and Foissner (1988) for accounts of this genus.

305

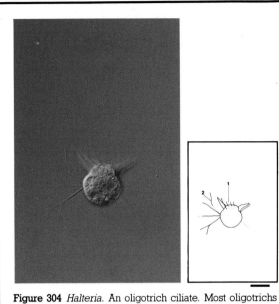

Figure 304 *Halteria*. An oligotrich ciliate. Most oligotrichs (see Figs 365–370) use the adoral zone of membranelles for feeding and locomotion. *Halteria* not only has an AZM (1) but also an equatorial girdle of stiff cirri (2), the action of which propels the cell with a bounding motion. This is a sufficiently distinctive trait for the genus to be identified from motion alone. Sometimes, the cell may come to rest with one pole attached to the slide or coverslip, under which circumstances it may look like a heliozoon. *Differential interference contrast.*

B   Without the bounding motion. GO TO 156

A   Cells with a marked torsion, the front twisted or with a spiral flange running along the body. Sometimes with spikes protruding posteriorly. GO TO 157 **156 (155)**

B   Not as 156A. GO TO 158

A Medusoid or mushroom-shaped, often with the posterior end drawn out as a spike. With an adoral zone of membranelles (AZM) following the spiral edge of the body, but with the rest of the body predominantly without cilia. 25–100 µm long.

Figs 306 & 307 CAENOMORPHA

See Jankowski (1964) and Fernandez-Galiano (1980) for comments on this heterotrich, which is related to *Metopus* (Step 159), and is usually found in anoxic sites.

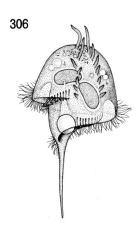

306

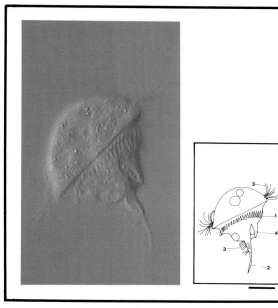

Figure 307 *Caenomorpha*. An umbrella-shaped heterotrich ciliate, found mostly in anoxic habitats. The adoral zone of membranelles (1) follows a spiral path, leading to the mouth (4), and ending at the base of the 'tail' (2). The somatic (locomotor) ciliature is reduced to a band (3) which lies adjacent to the membranelles. *Caenomorpha* typically has a cluster of refractile granules at the anterior end of the cell. *Differential interference contrast.*

B An elongate cell, but with the front end twisted. An adoral zone of membranelles (AZM) follows a spiral path, sometimes extending along much of the body. Kineties also follow a spiral path. Often with a dense granular mass at the front end. 50–150 µm long.

Figs 308 & 309 METOPUS

Both *Metopus* and *Caenomorpha* are most commonly encountered in organically enriched, usually anoxic, habitats (Jankowski, 1964). Such organisms may be referred to as sapropelic (Villeneuve-Brachon, 1940; Foissner, 1980a).

308

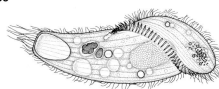

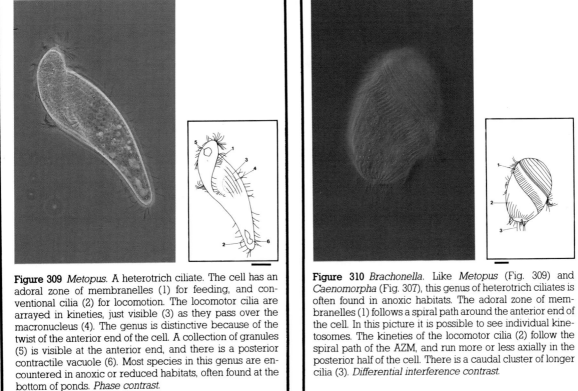

Figure 309 *Metopus*. A heterotrich ciliate. The cell has an adoral zone of membranelles (1) for feeding, and conventional cilia (2) for locomotion. The locomotor cilia are arrayed in kineties, just visible (3) as they pass over the macronucleus (4). The genus is distinctive because of the twist of the anterior end of the cell. A collection of granules (5) is visible at the anterior end, and there is a posterior contractile vacuole (6). Most species in this genus are encountered in anoxic or reduced habitats, often found at the bottom of ponds. *Phase contrast.*

Figure 310 *Brachonella*. Like *Metopus* (Fig. 309) and *Caenomorpha* (Fig. 307), this genus of heterotrich ciliates is often found in anoxic habitats. The adoral zone of membranelles (1) follows a spiral path around the anterior end of the cell. In this picture it is possible to see individual kinetosomes. The kineties of the locomotor cilia (2) follow the spiral path of the AZM, and run more or less axially in the posterior half of the cell. There is a caudal cluster of longer cilia (3). *Differential interference contrast.*

A Flattened and rigid cells, usually wedge-shaped. The body may be drawn out into folds and/or spikes, with the somatic cilia reduced to a few tufts. Usually from putrid waters.   Fig. 311 EPALXIS

An odontostome ciliate (Jankowski, 1964), *Epalxis* is related to *Epalxella* (Fig. 312), *Discomorphella* (Fig. 313), *Saprodinium* and *Myelostoma* (p. 188; Schrenk and Bardele, 1991).

311

B Cells not flattened or, if flattened, not rigid.   GO TO 159

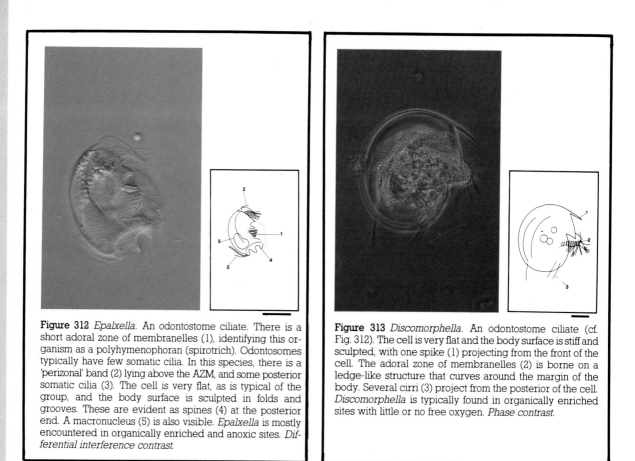

Figure 312 *Epalxella*. An odontostome ciliate. There is a short adoral zone of membranelles (1), identifying this organism as a polyhymenophoran (spirotrich). Odontosomes typically have few somatic cilia. In this species, there is a 'perizonal' band (2) lying above the AZM, and some posterior somatic cilia (3). The cell is very flat, as is typical of the group, and the body surface is sculpted in folds and grooves. These are evident as spines (4) at the posterior end. A macronucleus (5) is also visible. *Epalxella* is mostly encountered in organically enriched and anoxic sites. *Differential interference contrast.*

Figure 313 *Discomorphella*. An odontostome ciliate (cf. Fig. 312). The cell is very flat and the body surface is stiff and sculpted, with one spike (1) projecting from the front of the cell. The adoral zone of membranelles (2) is borne on a ledge-like structure that curves around the margin of the body. Several cirri (3) project from the posterior of the cell. *Discomorphella* is typically found in organically enriched sites with little or no free oxygen. *Phase contrast.*

159
(156)

A Scoop-shaped cells, planktonic.

GO TO 160

B Not scoop-shaped.

GO TO 161

160
(159)

A With a cylinder of rods (collectively referred to as a nasse or a basket) lying just internal to the mouth. Cell about 100 µm long.

Figs 314 & 315 PHASCOLODON

Ultrastructure is described by Tucker (1972). See also Foissner (1979b).

314

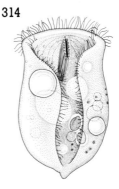

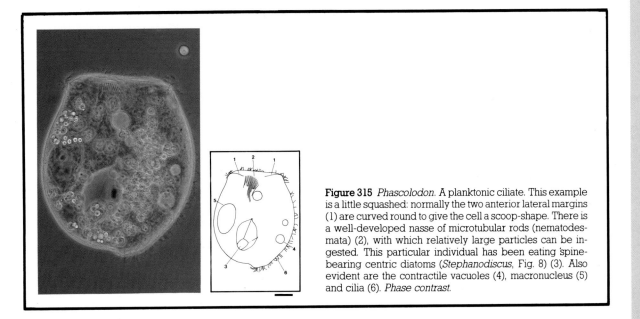

Figure 315 *Phascolodon*. A planktonic ciliate. This example is a little squashed: normally the two anterior lateral margins (1) are curved round to give the cell a scoop-shape. There is a well-developed nasse of microtubular rods (nematodesmata) (2), with which relatively large particles can be ingested. This particular individual has been eating spine-bearing centric diatoms (*Stephanodiscus*, Fig. 8) (3). Also evident are the contractile vacuoles (4), macronucleus (5) and cilia (6). *Phase contrast.*

B With what appears to be a band of ciliary membranelles leading into a crescent-shaped mouth cavity. Cell up to 1000 µm long.

Figs 316 & 317 BURSARIA

For a long time *Bursaria* was held to be a heterotrich (see notes after Step 161), but it is probably related to *Colpoda*. Like *Bursaria*, *Lembadion* (Figs 318 & 319) is shaped like a scoop, and is also encountered in the water column. It is a scuticociliate (Step 169).

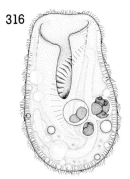

316

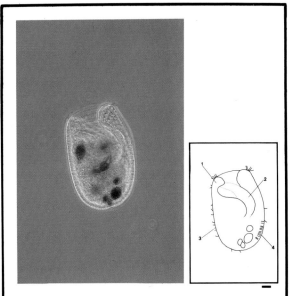

Figure 317 *Bursaria*. A planktonic ciliate that is related to *Colpoda* (Fig. 328). The organism is shaped like *Phascolodon* (Fig. 315), another planktonic ciliate. In both, the anterior margin (1) of the cell is drawn out, giving the entire cell a scoop-shape, and enabling it to drive quite large particles of food (in this individual, dinoflagellates and diatoms) into the buccal channel (2). Also visible are cilia (3) and extrusomes (4). *Phase contrast.*

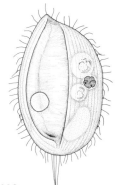

318

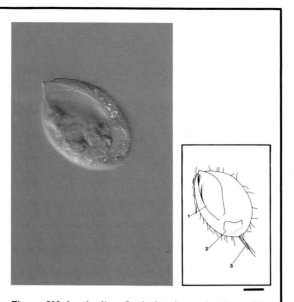

Figure 319 *Lembadion*. A planktonic scuticociliate. Like some other planktonic ciliates, such as *Phascolodon* (Fig. 315) and *Bursaria* (Fig. 317), the cell is scoop-shaped. The opening of the scoop (1) is lateral, rather than anterior, but the cell spirals on its longitudinal axis as it swims, and this helps to direct currents of water into the mouth. There is a well-developed undulating membrane. The contractile vacuole (2) and long caudal cilia (3) are also evident. *Differential interference contrast.*

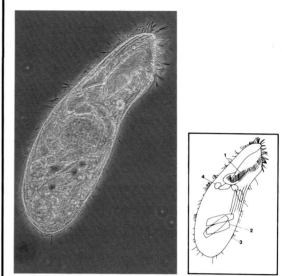

Figure 320 *Condylostoma*. A flattened, browsing heterotrich ciliate. The adoral zone of membranelles (1) extends across the front of the cell, curving sharply into the buccal cavity. A veil-like undulating membrane is also present, but it cannot be seen here. This particular cell has been feeding on diatoms (2). The cell has conventional somatic cilia (3) for locomotion. Also visible are some parts of the beaded macronucleus (4). *Phase contrast.*

161
(159)

A Cells without an adoral zone of membranelles (AZM). GO TO 165

B Cells that have an even covering of body cilia, in addition to an AZM.

THE HETEROTRICHS GO TO 162

The heterotrichs include some of the largest, most colourful, and architecturally most impressive ciliates. Because of the AZM (Fig. 320), they and the hypotrichs (see notes after Step 136) have been grouped together within the spirotrichs, and more recently within the Polyhymenophora. The Polyhymenophora were often regarded as the crown of the ciliate evolutionary tree, but recent work suggests that the heterotrichs may have diverged from ancestral stock at an early stage (Small and Lynn, 1985). Heterotrich cell bodies are often relatively large and flexible. They are distinguishable from hypotrichs because there are kineties on the body surface, but are most easily confused with some of the hypotrichs that have flexible bodies and several rows of ventral cirri. Some genera have already been keyed out (Steps 136 & 157).

162
(161)

A Bodies contractile. GO TO 163

B Bodies not contractile. GO TO 164

A Worm-like cells (up to 1 mm long). With the contractile vacuole located at the posterior end and a canal running towards the anterior of the cell.

Figs 321–323 SPIROSTOMUM

Isquith and Repak (1974) provide a key to species. *Spirostomum* is often found in polluted sites, or in sites with little or no oxygen.

321

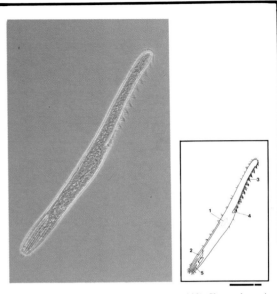

Figure 322 *Spirostomum* (see also Fig. 323). Shown here is *S. teres*, one of the smallest species in the genus. It is distinguished by the compact macronucleus (1). The cell is cylindrical, with a posterior contractile vacuole (2), and an adoral zone of membranelles (3) leading from the anterior of the cell to the cytostome (4). Each membranelle beats slightly out of synchrony with its neighbour, with what is called metachronal co-ordination. As a result, a wave-like pattern of activity passes along the organelles. This occurs so fast that it is difficult to see it without photography. The surface of the body is ridged with the kineties of locomotor cilia (5). *Phase contrast.* (Scale bar 100 μm.)

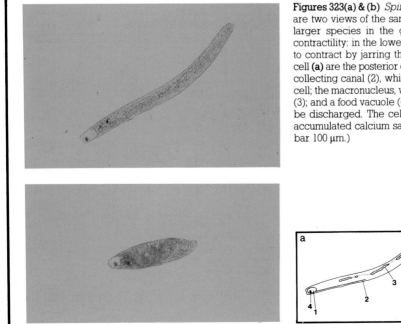

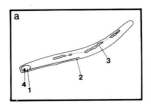

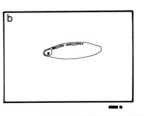

Figures 323(a) & (b) *Spirostomum* (see also Fig. 322). These are two views of the same cell of *S. ambiguum*, one of the larger species in the genus. The photographs illustrate contractility: in the lower view (**b**), the cell has been made to contract by jarring the glass slide. Visible in the upper cell (**a**) are the posterior contractile vacuole (1); its associated collecting canal (2), which extends towards the front of the cell; the macronucleus, which looks like a string of sausages (3); and a food vacuole (4) lying near the cytoproct ready to be discharged. The cells are often very opaque owing to accumulated calcium salts. *Bright field, green filter.* (Scale bar 100 μm.)

B Cone-shaped cells.

Stentor has the adoral zone of membranelles (AZM) at the broad end of the cone, but it may be difficult to see if the cells are contracted. They are large (up to 1 mm), with the contractile vacuole located near the anterior end of the cell. Many species are coloured (green, blue, pink). They may also attach to the substrate by their posterior ends and adopt a trumpet shape (see Step 118).

164
(162)

A Pink or red cells with the contractile vacuole at the posterior end. Most species have a well-developed undulating membrane alongside the adoral zone of membranelles (AZM). Body 50–350 μm long.

Figs 324–326 BLEPHARISMA

See Giese (1973) for an account of the biology of *Blepharisma*, and Hirschfield *et al.* (1965) and Larsen and Nilsson (1983) for comments on it and related genera. The cells usually eat bacteria, but they may become cannibalistic, and will then display dark red food vacuoles containing the partly digested residues of their confederates.

324

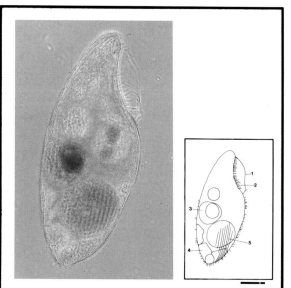

Figure 325 *Blepharisma*. Most members of this genus of heterotrich ciliates are pink. The mouth ciliature extends from the front of the cell to the cytostome (about a third of the way down the cell). The feeding ciliature includes the veil-like undulating membrane (1) and an adoral zone of membranelles (2). This species mostly feeds on bacteria and small protists, but it may occasionally become cannibalistic; the tell-tale pink remains of another cell of the same species are evident in one food vacuole (3). The contractile vacuole complex (4) is at the posterior end of the cell, and in this region the kineties (5) or lines of cilia are evident. *Phase contrast.* (Scale bar 100 μm.)

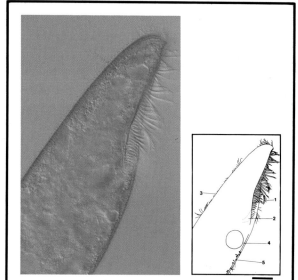

Figure 326 *Blepharisma*. Detail of the anterior end of a cell. The adoral zone of membranelles (1) can be seen leading from the anterior end of the cell to the cytostome (2). The membranelles are blocks of adhering cilia, and comparison may be made with individual cilia on the surface of the body (3). The body is highly vacuolated; food vacuoles (4) can be recognized because they contain bacteria. The cortex (5), which appears as a discrete layer at the surface of the cell, looks granular because of the small surface vesicles that contain the pink pigment. *Differential interference contrast.*

B Cells with anterior torsion, and the membranelles lying in a spiral groove. Species vary in size from 40–300 μm.
Step 157, Figs 308–310 METOPUS

A With evenly distributed body cilia.
GO TO 166

B Body cilia absent or patchily distributed.
GO TO 188

A Cells with cilia clustered in or around a buccal depression; these cilia generate currents of water from which particles of food are taken.
GO TO 167

Buccal (feeding) cilia may be located entirely within the buccal cavity, or extend onto the body surface. Feeding behaviour is often only apparent when the cells stop moving. The type of food cán be inferred from undigested coccoid bacteria or other small particles in food vacuoles.

B Not as 166A.
GO TO 173

A Flattened kidney-shaped cells, often found in soil, with the mouth located in a lateral invagination. The kineties converge on this region. With the contractile vacuole at the posterior end.
Figs 327 & 328 COLPODA

See Lynn (1976a & b, 1978), Novotny *et al.* (1977), Lynn and Malcolm (1983), Foissner (1980, 1987 *inter alia*) for the diversity of *Colpoda*, and Bardele (1983) for ultrastructural accounts of various members of the genus. Related to the colpodids are *Bursaria* (Figs 316 & 317), and *Cyrtolophosis* (Fig. 329), a genus that may be found attached to detritus in a mucus sheath.

327

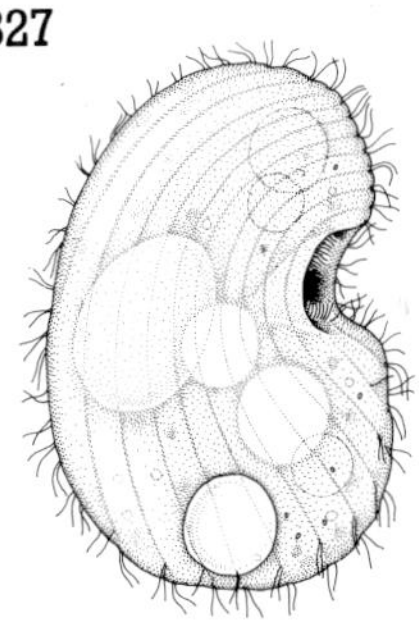

ALL SCALE BARS 20 μm UNLESS OTHERWISE INDICATED

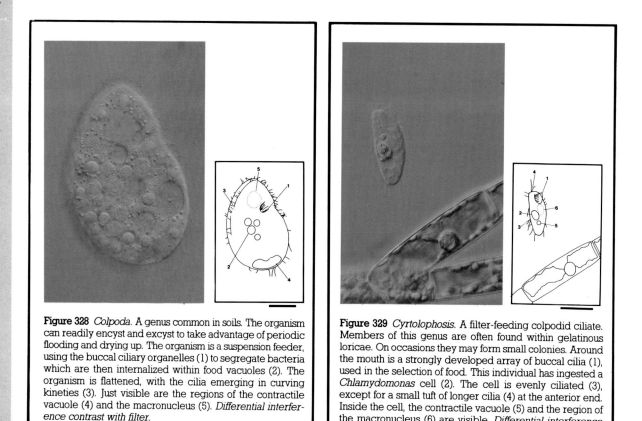

Figure 328 *Colpoda*. A genus common in soils. The organism can readily encyst and excyst to take advantage of periodic flooding and drying up. The organism is a suspension feeder, using the buccal ciliary organelles (1) to segregate bacteria which are then internalized within food vacuoles (2). The organism is flattened, with the cilia emerging in curving kineties (3). Just visible are the regions of the contractile vacuole (4) and the macronucleus (5). *Differential interference contrast with filter.*

Figure 329 *Cyrtolophosis*. A filter-feeding colpodid ciliate. Members of this genus are often found within gelatinous loricae. On occasions they may form small colonies. Around the mouth is a strongly developed array of buccal cilia (1), used in the selection of food. This individual has ingested a *Chlamydomonas* cell (2). The cell is evenly ciliated (3), except for a small tuft of longer cilia (4) at the anterior end. Inside the cell, the contractile vacuole (5) and the region of the macronucleus (6) are visible. *Differential interference contrast.*

B Not as 167A.   GO TO 168

168
(167)

A Ovoid cells that dart about, but frequently stop moving in order to feed. When feeding, a flat sheet of cilia is extended from one side of the body.   **THE SCUTICOCILIATES** GO TO 169

Scuticociliates are usually characterized by a large well-developed undulating membrane. Most are free-swimming organisms, but the relatively uncommon *Calyptotricha* (Wilbert and Foissner, 1980) (Fig. 330) is attached to detritus by means of a lorica that is open at both ends.

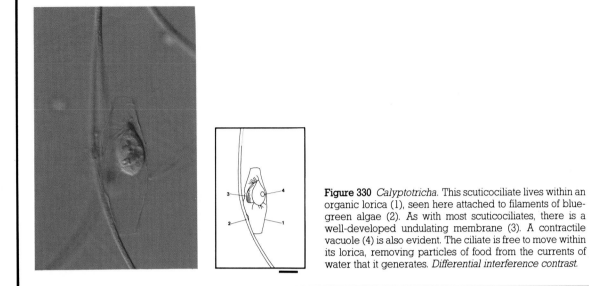

Figure 330 *Calyptotricha*. This scuticociliate lives within an organic lorica (1), seen here attached to filaments of blue-green algae (2). As with most scuticociliates, there is a well-developed undulating membrane (3). A contractile vacuole (4) is also evident. The ciliate is free to move within its lorica, removing particles of food from the currents of water that it generates. *Differential interference contrast.*

B Not as 168A.

GO TO 170

A Small cells (usually less than 30 μm) with relatively few somatic cilia.

Figs 331 & 332 CYCLIDIUM

See Didier and Wilbert (1981), Berger and Thompson (1960) and Bardele (1983) for accounts of the members and the fine structure of this genus. Common and widespread.

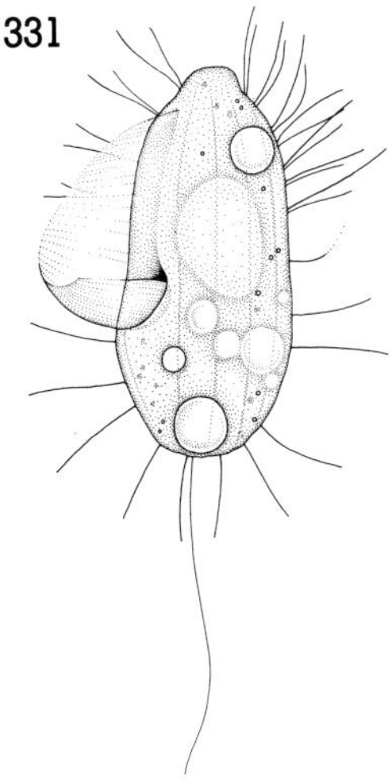

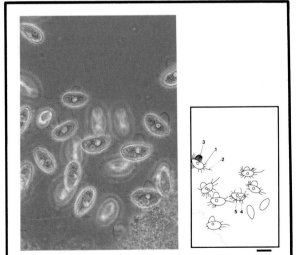

Figure 332 *Cyclidium*. A common suspension-feeding scuticociliate. The somatic cilia (1) are relatively sparse and long, and there is a single longer caudal cilium (2). When these ciliates feed, they become motionless and unfurl a sail-like undulating membrane (3). The cilia of the undulating membrane (barely visible here) are used to strain small suspended particles from the currents of water. The food is then packaged in food vacuoles (4) inside the cell. All cells have a contractile vacuole (5) which in the most common species is found at the posterior end. Here, however, it is located in the centre of the cell. *Phase contrast.*

B Cells larger than 30 μm in length, with a densely packed layer of long somatic cilia.

Figs 333 & 334 **PLEURONEMA**

For discussions of *Pleuronema*, see Dragesco (1968), and Grolière and Detcheva (1974).

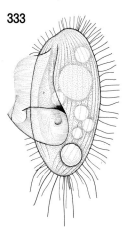

333

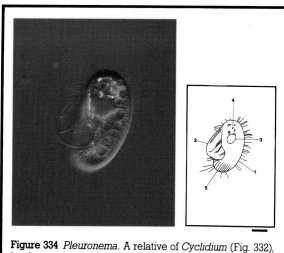

Figure 334 *Pleuronema*. A relative of *Cyclidium* (Fig. 332), but larger. Like its smaller relative, the cell has relatively long somatic cilia (1) which are spread out when the cell comes to rest in order to feed. A well-developed undulating membrane (2) is extended during feeding. Inside the cell, the macronucleus (3) and refractile crystals (4) are evident. The kineties may be seen as folds on the surface of the cell. *Differential interference contrast.*

A Thin, elongate cells in which some of the feeding cilia are curved, forming a passage from the anterior of the cell to the mouth, and within which the undulating membrane beats. Body 50–100μm long.

Figs 335 & 336 **COHNILEMBUS**

See Didier and Detcheva (1974) for observations on *Cohnilembus*.

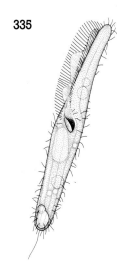

335

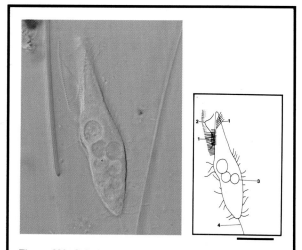

Figure 336 *Cohnilembus*. The most obvious feature of this ciliate is the highly developed buccal ciliature which stretches from the anterior end of the cell to a mouth about halfway down. Two lines of cilia (or a single line of U-shaped cilia) (1) form a channel within which a single line of cilia undulates (2). This enables small particles of food to be segregated from the surrounding medium, channelled towards the cytostome, and then packaged within food vacuoles (3). The body is evenly ciliated and there is a long caudal cilium (4). *Differential interference contrast.*

B Not as 170A. GO TO 171

A The mouth is at the base of a short groove located just below a slightly jutting or twisted anterior part of the body. With a central contractile vacuole. Body 40–100 μm. Figs 337–339 COLPIDIUM

Colpidium is common in organically enriched sites. Foissner and Schiffmann (1980) give an account of the genus, and other studies have been made by Jankowski (1967a), Lynn and Didier (1978), and Iftode *et al.* (1984).

337

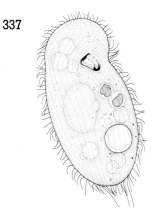

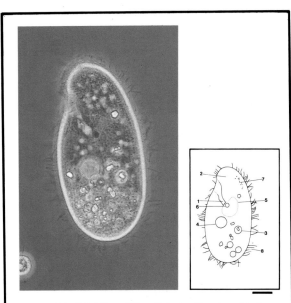

Figure 338 *Colpidium*. This ciliate is often found in large numbers in organically enriched and slightly anoxic sites. The mouth (1), which lies below a slightly protruding anterior part of the body (2), incorporates ciliary organelles that are used to concentrate suspended particles and deliver them for packaging into food vacuoles (3). The large, empty-looking vacuole (4) near the centre of the cell is the contractile vacuole. Also visible within the cell are the regions of the macronucleus (5), the micronucleus (6) and mitochondria (7). The cell is evenly ciliated (8). *Phase contrast, artificial light film.*

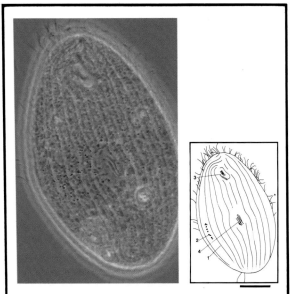

Figure 339 *Colpidium* dividing. Illustrated is a slightly earlier stage in division than that shown for *Tetrahymena* (Fig. 341). The photograph is of a slightly squashed cell with the surface in focus. The kineties can be seen as light lines (1), around which, inside the cell, are clustered mitochondria (2). Before the cell divides, the number of somatic cilia needs to double, and the new mouth of the posterior daughter cell (opisthe) must form. Here, the well-formed mouth of the presumptive anterior daughter cell (the proter) may be seen (3), along with some of the compound feeding ciliary organelles. The posterior mouthparts (4) lie on the surface of the cell, as the mouth cavity has not yet formed. *Phase contrast.*

B Without anterior protrusion. GO TO 172

172
(171)

A Small (less than 50 μm), pear-shaped cells with a small anterior mouth, and the contractile vacuole in the posterior part of the cell. Sometimes with a caudal cilium. Often associated with waters that contain many animals, or with dead and/or decaying animal matter.
Figs 340–342 **TETRAHYMENA**

Tetrahymena has been widely exploited in physiological and biochemical studies (Eliott, 1973). Since Nanney and McCoy (1977) revised the criteria for identifying species, many have been added to the genus (Batson, 1983; Williams *et al.*, 1984). Biochemical techniques are now required for identification of many species (Corliss and Daggett, 1983).

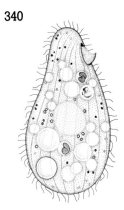

340

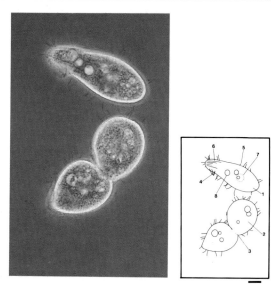

Figure 341 *Tetrahymena* (see also Fig. 342). Probably the most studied genus of all the ciliates, as it can now be grown in bacteria-free cultures. As a result, it has become widely used for experimental studies on its physiology, biochemistry, etc.. Two cells are shown: a normal trophic cell (1); and a dividing cell (2) with its transverse furrow (3). In the wild, *Tetrahymena* feeds mostly on suspended bacteria, often in association with decaying animal material. Bacteria are concentrated by the buccal ciliature around the mouth (4). The cell body is evenly ciliated (5). Also evident in the normal cell are slight furrows (kineties) (6) on the surface, the macronucleus (7) and the fluid-filled food vacuoles (8). *Phase contrast.*

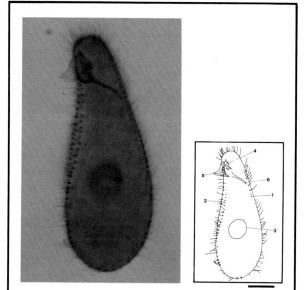

Figure 342 *Tetrahymena*. This is a fixed preparation that has been stained with protargol. This silver stain reveals basic (negatively charged) proteins, such as tubulins and histones. For this reason, basal bodies (1), cilia (2) and the macronucleus (with histones) (3) can be seen. The photograph has been taken with a shallow focal depth, and is in focus through the centre of the cell. Several rows (kineties) of kinetosomes are visible at the margins of the cell, following the longitudinal axis of the body. Around the mouth region, the densely packed kinetosomes supporting the cilia of the three membranelles (4) and the curving undulating membrane (5) have stained clearly. Also revealed are the microtubular ribbons of the cytopharynx (6), leading from the mouth into the cytoplasm of the cell. *Bright field.*

B Cells with a layer of trichocysts under the body surface. With two contractile vacuoles, typically with radiating collecting canals. (One species (Figs 347–348) has a different kind of contractile vacuole complex.) With a caudal tuft of cilia, and a mouth that lies near the middle of the body, at the posterior end of a pre-oral groove. Either elongate or flattened ovoid (foot-shaped).   Figs 343–359 **PARAMECIUM**

There are many general accounts of this very common and familiar genus of ciliate: Wichterman (1953, 1985), Wagtendonk (1974) and Görtz (1988) all give general accounts including outlines of the composition of the genus.

As with *Tetrahymena*, some species can only be identified by their biochemical characteristics (Corliss and Daggett, 1983). Morphological species can be distinguished by their shape, being either elongate (slipper-shaped) (Fig. 343) or foot-shaped (Figs 348–359), and by the form of the micronuclei (Fig. 357). One common species (*P. bursaria*) contains endosymbiotic algae (Figs 349–358).

These are peniculine ciliates, the closest relatives of which are *Frontonia* (Step 150), *Urocentrum* (Figs 387 & 388) and *Neobursaridium* (Fig. 360).

343

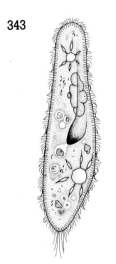

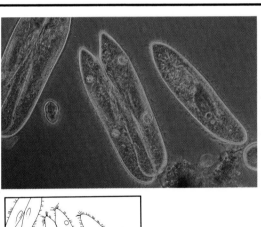

Figure 344 *Paramecium*. Both a typical individual and conjugating cells (1) are shown. The single cell contains a macronucleus (2) and a spherical adpressing micronucleus (3). This particular nuclear configuration, combined with the shape of the cell, identifies the organism as being *P. caudatum*. Conjugation is a mechanism that allows the exchange of genetic information between mating cells. The cells become joined, a cytoplasmic channel forms between them, and a gametic nucleus (product of meiosis of the micronucleus) is exchanged. Contractile vacuoles (4) are also visible. *Phase contrast*. (Scale bar 100 μm.)

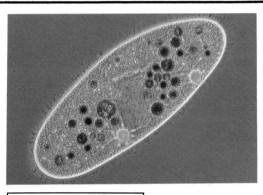

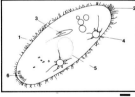

Figure 345 *Paramecium* with food vacuoles. This slightly squashed cell has been fed with bacteria stained black and red, in order to show the food vacuoles. The cell is evenly covered with cilia (1) but, as with all *Paramecia*, there is a caudal tuft (2) of longer cilia. The mouth may be seen as a channel (3) with a densely packed line of cilia. Food is pushed down the channel, ultimately to be packed into the food vacuoles (coloured). Also visible in the cell are the two contractile vacuoles with their radiating collecting canals (4), the macronucleus (5) and trichocysts (6). *Phase contrast*.

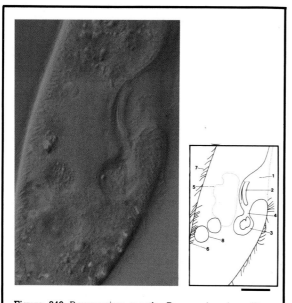

347

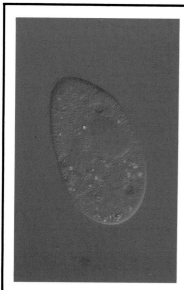

Figure 346 *Paramecium* mouth. *Paramecium* is a filter-feeding ciliate. The body is shaped to create a channel from the front end of the cell to the buccal cavity, seen here (1) in a side view. In the buccal cavity lie the compound ciliary organelles (2) (membranelles or peniculi) that create the currents of water from which particles are taken. The parallel lines of cilia that make up the peniculi are just discernible. Food is packed into food vacuoles (3) which form at the cytostome (4) at the base of the buccal cavity. A macronucleus (5), trichocysts (6), cilia (7) and a contractile vacuole (8) are also evident. *Differential interference contrast.*

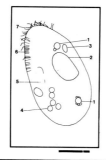

Figure 348 *Paramecium putrinum.* The smallest species in the genus. This species (Fig. 347) is also called *P. trichium.* Unlike other members of the genus, the contractile vacuoles (1) do not have radiating collecting canals. There is one large macronucleus (2) and an adjacent compact micronucleus (3). Food vacuoles (4) can be distinguished because they contain densely packed bacteria. Also visible are the mouth (5), trichocysts (6) and cilia (7). *Differential interference contrast.* (Scale bar 100 µm.)

Figures 349–358 This series of photographs of *Paramecium bursaria* attempts to illustrate the different contrast enhancement techniques that can be used with free-living protozoa. Various characteristics of this species may be seen in Fig. 352. As with other species in this genus (see, e.g. Fig. 345), the cell is evenly covered with cilia (1), but there is also a caudal tuft of longer cilia (2). The mouth (3) appears as a narrow cavity containing densely packed cilia. Food vacuoles (4) are rare, as the cell contains many symbiotic green algae which provide energy in the form of released by-products of photosynthesis. Two contractile vacuoles (5), trichocysts (6) and the macronucleus (7) may also be seen, although the macronucleus is clearer in Fig. 357 than in Fig. 352.

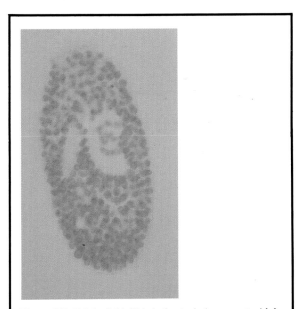

Figure 349 *Bright field.* This is the technique most widely advocated in textbooks and schools, etc.. Natural colour differences are evident, but fine details cannot be seen.

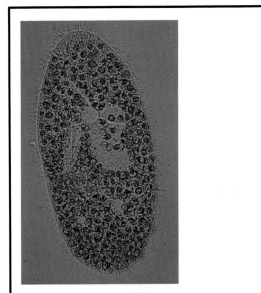

Figure 350 *Bright field, with the condenser iris closed.* The simplest way to generate sufficient contrast for most of the parts of the cell to be seen. The same effect can be achieved by lowering the condenser, or by moving it out of alignment with the optical axis of the microscope. This technique is often frowned upon by the purists because they (rightly) argue that resolution is lost. However, the technique is justified because visibility is often more important than resolution.

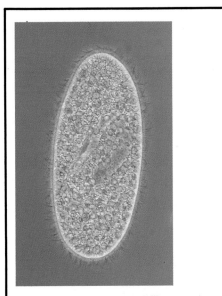

Figure 351 *Phase contrast.* Differences in optical density are revealed as regions of darkness and light. Cilia can be clearly seen against the surrounding fluid. Some degree of optical confusion may occur when one object lies on top of another, e.g. deep in the cytoplasm. However, it is a widely used and effective technique, especially for smaller protozoa.

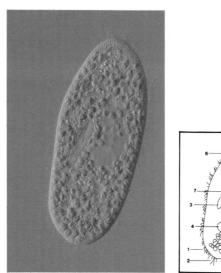
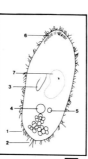

Figure 352 *Differential interference contrast (Nomarski).* Gradients (e.g. between an organelle and the cytoplasm) in refractive index are shown up as a shadow effect. Cilia are seen as a dark and light shadow against the background, producing a very appealing three-dimensional effect. A special advantage of the technique is that a very thin optical slice is taken through the specimen. Consequently, there is very little optical confusion, and a clean, crisp image can be obtained. However, this aspect of the technique also has its disadvantages, as only those pieces of the cell that lie in a single plane are visible.

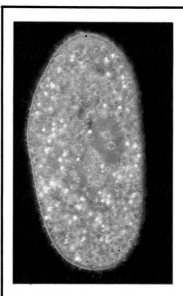

Figure 353 *Dark ground.* The specimen is illuminated very obliquely, so that no light can pass directly from the light source into the objective of the microscope. In the absence of an object, no light can be observed. However, an object in the light path will refract the light and appear bright against a dark background. The technique is aesthetically very appealing. It may also be used to show up very small objects. An opaque disc on the top of the condenser can generate this effect, as may the use of the wrong size of phase contrast rings in the condenser or lamp housing.

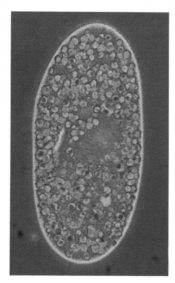

Figure 354 *Interference contrast.* Differences in refractive index are converted into different colours. Colours originally present in the object, such as the greenness of algae, are retained.

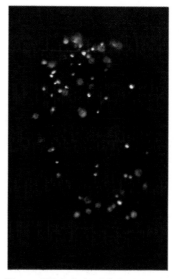

Figure 355 *Polarized light.* The object is illuminated with polarized light, and then viewed through a second polarizing filter at right angles. Crossed polarizers remove all light except that which has passed through crystalline structures. Consequently, this technique reveals crystals inside the cytoplasm. The crystals are also visible in previous micrographs, but are easily overlooked.

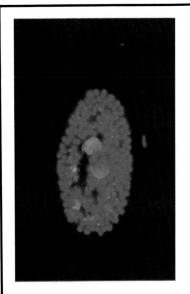

Figure 356 *Fluorescence microscopy (with DAPI stain).* The specimen is first exposed to DAPI, which stains DNA. The cell is then illuminated with a high energy light source (usually ultraviolet). DAPI and some natural molecules, such as the chlorophyll in the algae, will absorb radiation, become excited and re-emit radiation at a longer wavelength – in this case in the visible range. A filter is placed between the specimen and the eyepieces to remove the ultraviolet light, so that just the fluorescent colours are seen. Here, the algal chlorophyll fluoresces red, and the DNA in the nuclei fluoresces green.

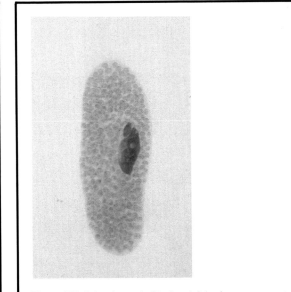

Figure 357 *Selective stain (Feulgen).* It is often necessary to fix protozoa cells and to stain them selectively in order to extract information not (easily) visible in living cells. With *Paramecium*, it is necessary to discover the arrangement of the micronucleus/i to identify species. This particular cell has been stained with Schiff's reagent, following the Feulgen technique, and this has shown up the macronucleus and the micronucleus of the ciliate; the tiny nuclei in each algal cell have also been stained. There are few examples of selective stains in this book; others are the silver techniques (Figs 255, 342 & 359).

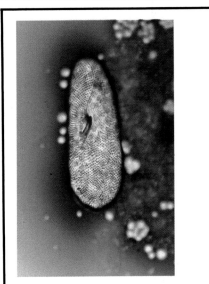

Figure 358 *Negative staining.* One of the simplest methods of preparing permanent preparations is to mix a very fine dye or stain with a sample, and then simply allow it to dry. Once dry, the stain settles into any irregularities on the surface (e.g. the mouth and the origin of the cilia), as well as settling around the preparation. The stain used here is nigrosin, but certain inks have the same effect (*N.B.* some inks are toxic to protozoa). Not all cells remain intact as they dry out.

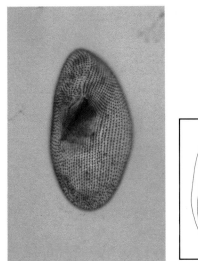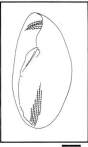

Figure 359 *Silver-staining.* An example of a Chatton-Lwoff-stained cell. Silver-staining is normally used to show up the bases of cilia, and the various techniques are needed for formal descriptions of ciliates. Each spot visible here is made up of three smaller dots (two ciliary bases and a parasomal sac). The technique reveals the kineties clearly; the closely packed cilia of the mouth have shown up almost black. Near the posterior end of the cell is the slit-like opening of the cytoproct (cell anus).

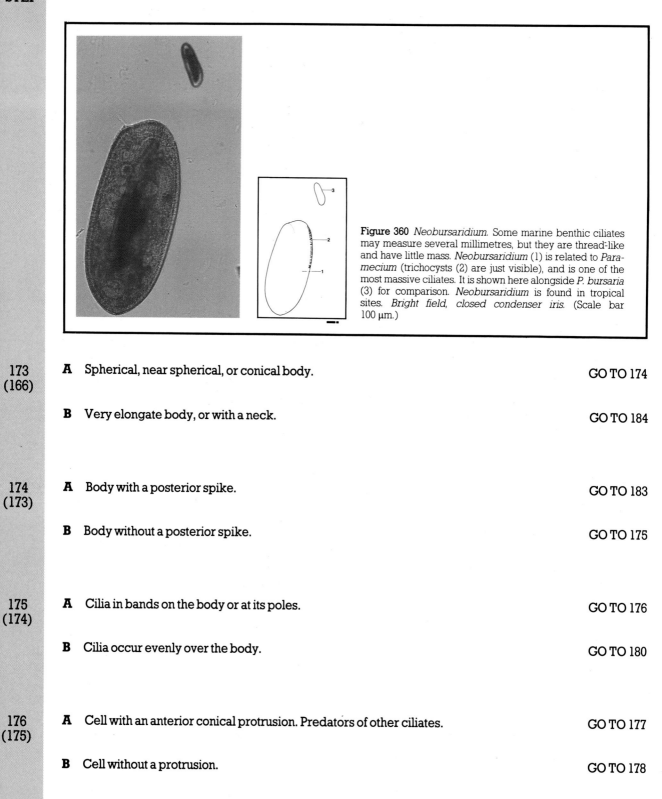

Figure 360 *Neobursaridium.* Some marine benthic ciliates may measure several millimetres, but they are thread-like and have little mass. *Neobursaridium* (1) is related to *Paramecium* (trichocysts (2) are just visible), and is one of the most massive ciliates. It is shown here alongside *P. bursaria* (3) for comparison. *Neobursaridium* is found in tropical sites. *Bright field, closed condenser iris.* (Scale bar 100 μm.)

**173
(166)**

A Spherical, near spherical, or conical body. GO TO 174

B Very elongate body, or with a neck. GO TO 184

**174
(173)**

A Body with a posterior spike. GO TO 183

B Body without a posterior spike. GO TO 175

**175
(174)**

A Cilia in bands on the body or at its poles. GO TO 176

B Cilia occur evenly over the body. GO TO 180

**176
(175)**

A Cell with an anterior conical protrusion. Predators of other ciliates. GO TO 177

B Cell without a protrusion. GO TO 178

A   With two bands of cilia: one anterior and one equatorial. Body 50–150 μm.   Figs 361 & 362 DIDINIUM

Wessenberg and Antipa (1968, 1970) provide a graphic account of the feeding behaviour and ultrastructure of *D. nasutum*, which feeds more or less exclusively on *Paramecium* species. General ultrastructural accounts are given by Rodrigues de Santa Rosa and Didier (1976), and Rieder (1971).

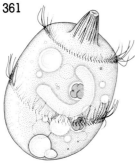

361

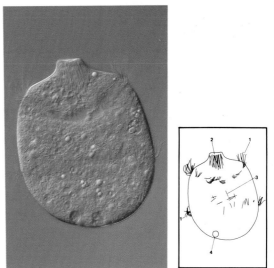

Figure 362 *Didinium*. A predatory ciliate that swims using two bands of cilia (1). There is an anterior cone which contains numerous extrusomes (2), used to impale and kill its prey, *Paramecium*. Some extrusomes (3) may be seen in the cytoplasm. Also visible is a posterior contractile vacuole complex (4). The ciliate in this picture is squashed (cf. Fig. 361). *Differential interference contrast.*

B   With a single band of cilia.

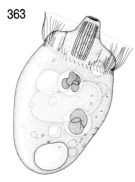

363

Figs 363 & 364 MONODINIUM

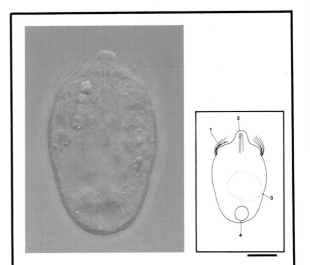

Figure 364 *Monodinium*. This genus is similar to *Didinium* (Fig. 362), but movement is achieved by a single band of cilia (1). As with *Didinium*, there is an anterior cone (2) which contains extrusomes, used in the capture of prey (ciliates). The diffuse area (3) inside the cell is the macronucleus. A posterior contractile vacuole is also evident (4). *Differential interference contrast.*

A With a strongly developed anterior adoral band of membranelles (AZM) around the anterior of the cell, and no other cilia. Body 30–100 μm.

Figs 365 & 366 STROBILIDIUM

Strobilidium cells are oligotrich ciliates. Like hypotrichs and heterotrichs (Steps 136 & 161), they have an AZM. Unlike the hypotrichs and heterotrichs, there are no somatic cilia and the AZM is used for motion as well as for food capture. One genus (*Halteria*) of oligotrichs has already been encountered (Step 155). *Strombidium* (Figs 367 & 368) looks very like *Strobilidium*, but differs because it has an organic lorica around the base of the cell. *Strobilidium* (Deroux, 1974) usually rotates near the substrate, as if tied to it by a thread of mucus; the thread is usually invisible. Some oligotrichs have a more substantial lorica, especially evident in the tintinnids, most of which are found in the sea. A few freshwater planktonic species do occur (Foissner and Wilbert, 1979; Figs 369 & 370). Oligotrichs are reviewed by Carey and Maeda (1985).

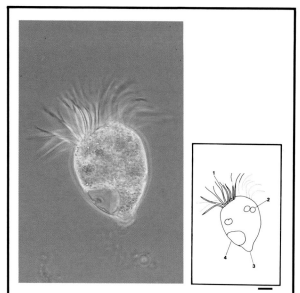

Figure 366 *Strobilidium*. An oligotrich ciliate. The only visible cilia are those that make up the adoral zone of membranelles (1) as it stretches around the front of the cell. Using the AZM, the organism can collect algal cells as food (2). The cell is often found apparently attached to the substrate by an invisible thread extending from the posterior end (3). It will spin and jerk at a fixed distance from the substrate. Occasionally it breaks free and can then move with great speed. A contractile vacuole (4) is visible. *Stobilidium* is most easily confused with *Strombidium* (Fig. 368).

365

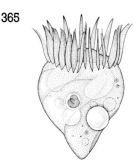

367

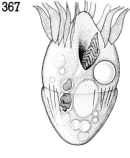

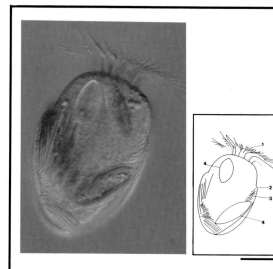

Figure 368 *Strombidium*. An oligotrich ciliate. The only visible cilia are those of the adoral zone of membranelles (1), used for feeding and locomotion. This genus is easily confused with *Strobilidium*. *Strombidium* may be distinguished by the presence of a lorica-like sheath, evidenced by a 'shoulder' (2) in the lateral profile; the genus also has 'trichites' (3). This particular specimen has been eating diatoms (4). *Phase contrast*.

369

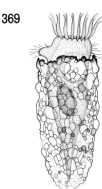

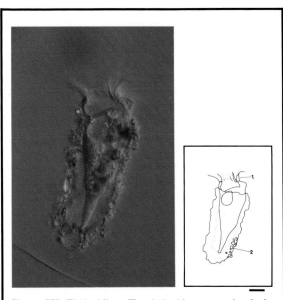

Figure 370 *Tintinnidium*. The tintinnids are mostly plank-tonic oligotrichs. The only visible cilia are those that make up the adoral zone of membranelles (1) as it stretches around the front of the cell. These cilia are used for feeding and locomotion. All tintinnids are enclosed by a lorica (2), which in this genus incorporates agglutinated material. Tintinnids are common in marine habitats. *Differential interference contrast.*

B Not as 178A.   GO TO 179

A With a band of cilia at the posterior and the anterior end. About 100 μm long.   179
(178)
                                          Fig. 371 TELOTROCHIDIUM

Telotrochidium may be difficult to distinguish from the mobile telotroch larvae of sessile peritrichs (Fig. 236), and from peritrich cells that have separated from their stalks and sprouted a temporary basal (trochal) band of cilia. For taxonomy, see Foissner (1976), and for general comments, see notes after Step 119. *Hastatella* (Figs 372 & 373) and *Astylozoon* (Figs 389 & 390) are related (Foissner, 1977).

371

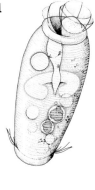

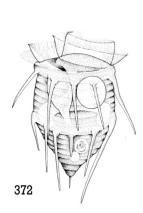

372

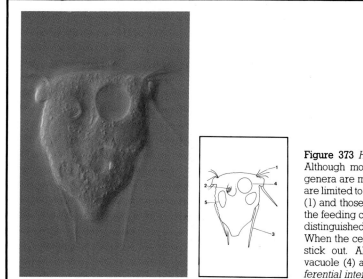

Figure 373 *Hastatella.* A free-swimming peritrich ciliate. Although most peritrichs are sessile, a small number of genera are motile. As in the majority of peritrichs, the cilia are limited to those around the broad anterior end of the cell (1) and those in the buccal cavity (2). The cilia that create the feeding current also pull the cell forward. This genus is distinguished by the cytoplasmic spines (3) on the body. When the cell stops and contracts, the spines are made to stick out. Also visible is an empty-looking contractile vacuole (4) and two profiles of the macronucleus (5). *Differential interference contrast.*

B   Small, with forked tentacles at the anterior end. Body 20–30 μm.

Figs 274 & 374 MESODINIUM

Mesodinium is a genus best known from marine habitats, where it is widespread and may have symbiotic (red) algae (Lindholm, 1986). It is occasionally found in freshwater habitats. When active, it moves with a jerky motion, and the tentacles may be withdrawn. It is distinguished from *Halteria* (see Step 155 above) by its size and by the absence of equatorial spikes. Small and Lynn (1985) use *Myrionecta* to house some species.

374

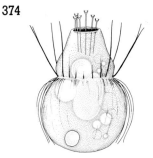

**180
(175)**

A   The body appears to be covered in a layer of plates, rather like a barrel. The mouth is located at the anterior end, and is used for feeding mainly on dead or dying animal tissue. Body (mostly) about 50 μm.

Figs 375–377 COLEPS

See Foissner (1984) for recent taxonomy of the better known species. For details of those species that have symbiotic algae, see Christopher and Patterson (1983), and Klaveness (1984). Ultrastructure is dealt with by Huttenlauch and Bardele (1987), and Lynn (1985).

375

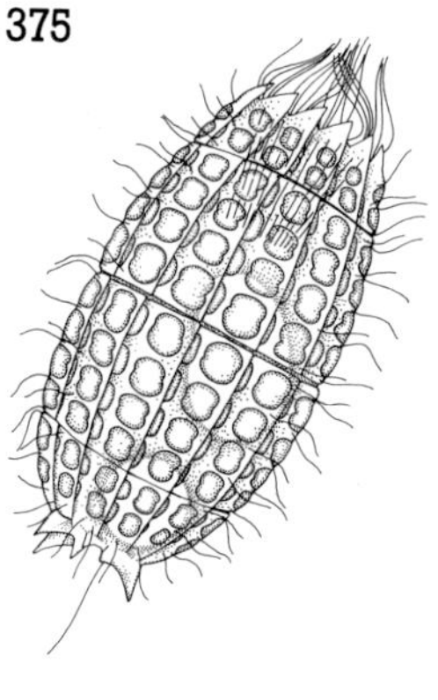

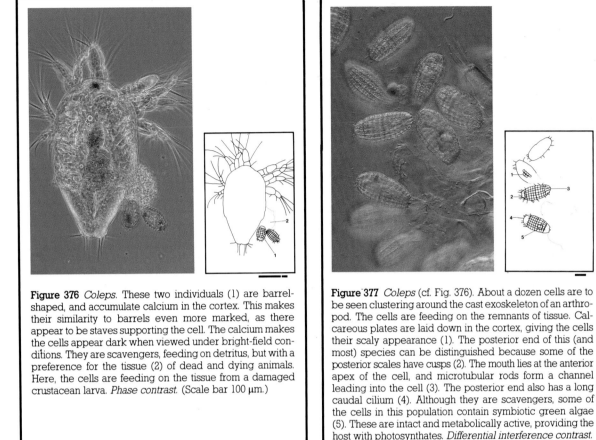

Figure 376 *Coleps*. These two individuals (1) are barrel-shaped, and accumulate calcium in the cortex. This makes their similarity to barrels even more marked, as there appear to be staves supporting the cell. The calcium makes the cells appear dark when viewed under bright-field conditions. They are scavengers, feeding on detritus, but with a preference for the tissue (2) of dead and dying animals. Here, the cells are feeding on the tissue from a damaged crustacean larva. *Phase contrast*. (Scale bar 100 μm.)

Figure 377 *Coleps* (cf. Fig. 376). About a dozen cells are to be seen clustering around the cast exoskeleton of an arthropod. The cells are feeding on the remnants of tissue. Calcareous plates are laid down in the cortex, giving the cells their scaly appearance (1). The posterior end of this (and most) species can be distinguished because some of the posterior scales have cusps (2). The mouth lies at the anterior apex of the cell, and microtubular rods form a channel leading into the cell (3). The posterior end also has a long caudal cilium (4). Although they are scavengers, some of the cells in this population contain symbiotic green algae (5). These are intact and metabolically active, providing the host with photosynthates. *Differential interference contrast*.

B Without plates. GO TO 181

A Mouth apical. GO TO 182

181
(180)

B The mouth is located away from the anterior pole, and is supported internally by a cylinder of stiff rods (basket or nasse). These organisms feed on algae, the remains of which can often be seen inside the cell. Body 30–300 μm, and circular or slightly flattened in cross section. Figs 378 & 379 NASSULA

Tucker (1978) gives accounts of the feeding behaviour, and Foissner (1979) deals with some taxonomy.

Drepanomonas (Fig. 380) is an unusually shaped relative of *Nassula*, typically found in sphagnum moss. It is also argued that the chonotrich ciliates, such as *Spirochona* (Fig. 381) from gills of *Gammarus*, are related to this group.

378

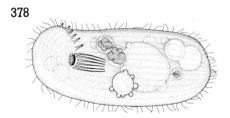

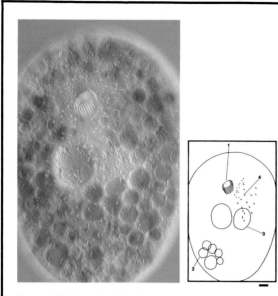

Figure 379 *Nassula*. This ciliate uses its well-developed basket or nasse (1) of microtubular rods to ingest blue-green algae. As digestion progresses, the photosynthetic pigments within the food vacuoles (2) are broken down, creating orange and lilac colours. After feeding, the whole ciliate may have an orange or pink colour. Also evident are a contractile vacuole (3) and extrusomes (4). The cell is round in cross section. *Differential interference contrast.*

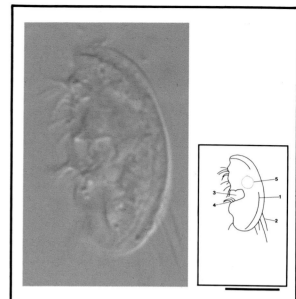

Figure 380 *Drepanomonas*. Encountered in various habitats, but especially in mosses, these cells are rigid, with the surface sculpted into curving folds (1). From the folds arise the somatic cilia (2). The indentation (3) on the concave side is the site of the cytostome. Several clusters of cilia (4) occur around this region. The genus is not very well known, and the full structure and feeding habits have yet to be elucidated. Also visible is the region containing the macronucleus (5). *Differential interference contrast.*

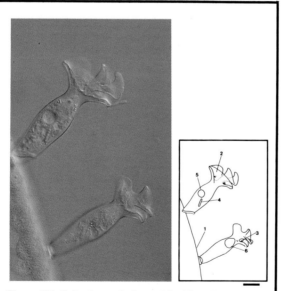

Figure 381 *Spirochona*. A chonotrich ciliate, found mostly on the surfaces of crustacea. (All chonotrich ciliates are ectosymbionts.) This species was photographed on the gill plate (1) of the freshwater louse, *Gammarus*. The anterior end of the cell has a spiral fold of cytoplasm (2), in the valley of which lie the cilia (3). One of these cells has been eating fragments of blue-green algal filaments (4). Other visible structures are the contractile vacuole (5) and the macronucleus (6). *Differential interference contrast.*

ALL SCALE BARS 20 μm UNLESS OTHERWISE INDICATED

A   Relatively small cells (about 20 µm), with long, tightly packed and slowly beating cilia. Sometimes with a posterior caudal cilium. Makes occasional jumps. Polar mouth, but without distinctive nematodesmata.

Figs 382 & 383 UROTRICHA

See Foissner (1979, 1983), Dragesco *et al.* (1974) for accounts of this genus.

382

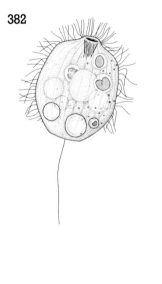

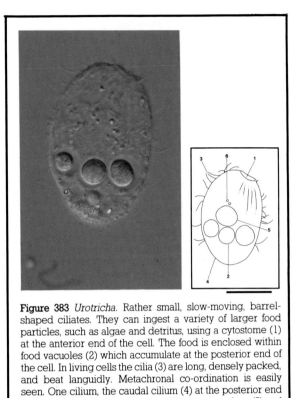

Figure 383 *Urotricha.* Rather small, slow-moving, barrel-shaped ciliates. They can ingest a variety of larger food particles, such as algae and detritus, using a cytostome (1) at the anterior end of the cell. The food is enclosed within food vacuoles (2) which accumulate at the posterior end of the cell. In living cells the cilia (3) are long, densely packed, and beat languidly. Metachronal co-ordination is easily seen. One cilium, the caudal cilium (4) at the posterior end of the cell, is longer than the rest. The macronucleus (5) and some mitochondria (6) are also visible. *Differential interference contrast.*

B   Medium to large cells (50–300 µm) in which the mouth is located at the anterior pole, and under it are extrusomes and/or nematodesmata.

Figs 384–386 PRORODON

384

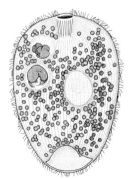

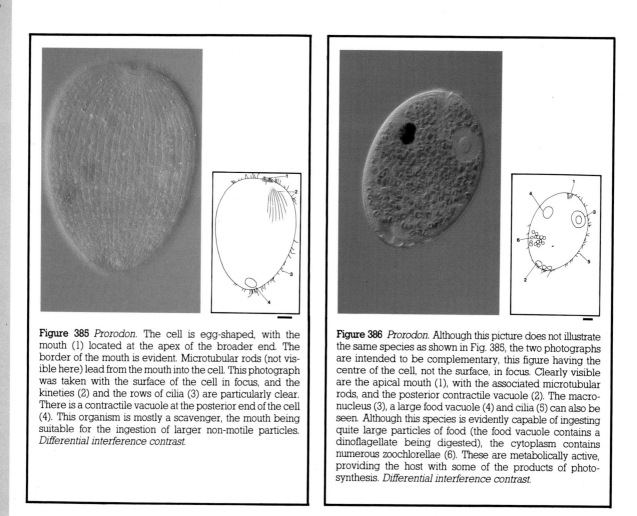

Figure 385 *Prorodon*. The cell is egg-shaped, with the mouth (1) located at the apex of the broader end. The border of the mouth is evident. Microtubular rods (not visible here) lead from the mouth into the cell. This photograph was taken with the surface of the cell in focus, and the kineties (2) and the rows of cilia (3) are particularly clear. There is a contractile vacuole at the posterior end of the cell (4). This organism is mostly a scavenger, the mouth being suitable for the ingestion of larger non-motile particles. *Differential interference contrast.*

Figure 386 *Prorodon*. Although this picture does not illustrate the same species as shown in Fig. 385, the two photographs are intended to be complementary, this figure having the centre of the cell, not the surface, in focus. Clearly visible are the apical mouth (1), with the associated microtubular rods, and the posterior contractile vacuole (2). The macronucleus (3), a large food vacuole (4) and cilia (5) can also be seen. Although this species is evidently capable of ingesting quite large particles of food (the food vacuole contains a dinoflagellate being digested), the cytoplasm contains numerous zoochlorellae (6). These are metabolically active, providing the host with some of the products of photosynthesis. *Differential interference contrast.*

183
(174)

A In addition to a spike, broad bands of cilia wrap around the wider upper part of the body, which is 60–100 μm long.

Figs 387 & 388 **UROCENTRUM**

The spike is a caudal tuft of cilia, which appears to produce the mucus used to connect cells loosely to the substrate. Attached cells seem to spin in one place. With a basal contractile vacuole, which has radiating collecting canals, and with trichocysts. Related to *Paramecium*. An ultrastructural study has been published by Didier (1970).

387

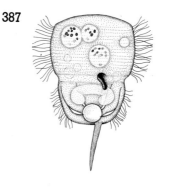

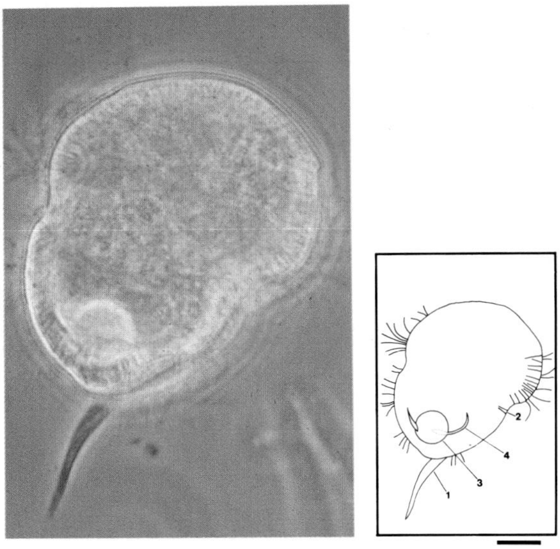

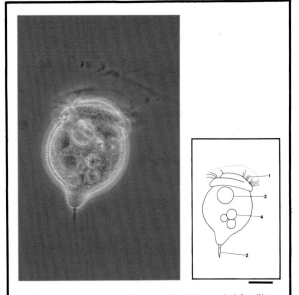

Figure 388 *Urocentrum*. This genus behaves a little like a spinning top: the tail (a caudal tuft of cilia) (1) loosely attaches to debris and the cell often appears to spin round. *Urocentrum* feeds mostly on suspended bacteria, and is related to *Paramecium*. Like most members of the latter genus, it has trichocysts (2) and a contractile vacuole (3) with radiating collecting canals (4). *Phase contrast.*

B With anterior wreaths of cilia, but not on the body. Most species about 50 μm long.

Figs 389 & 390 ASTYLOZOON

Astylozoon is a peritrich ciliate, the posterior spike of which is a reduced stalk. It is one of the few motile sessiline peritrichs (see notes after Steps 119 & 179). For taxonomy, see Foissner (1975, 1977).

389

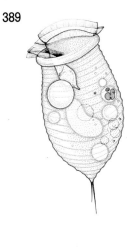

Figure 390 *Astylozoon*. A free-swimming peritrich ciliate. As with most peritrichs, feeding (and, in this case, movement) is achieved by means of the wreath of cilia around the broadened anterior end of the cell (1). The posterior end of the cell has a small spike (2). A contractile vacuole (3) and several food vacuoles (4) lie inside the cell. *Phase contrast.*

A Elongate body, often with a contractile neck (but not one that is clearly demarcated from the body). The mouth lies at the apex of the body.

Step 154, Figs 300–302 LACRYMARIA

184
(173)

B Neck clearly distinct from the body.

GO TO 185

A Slim, cigar-shaped body with a curved anterior extension that bears extrusomes. The mouth lies at the base of the extension where it joins the body. From 100 to more than 1000 µm.   Figs 391 & 392 DILEPTUS

Dileptus is reviewed by Dragesco (1963), and Grain and Golinska (1969).

391

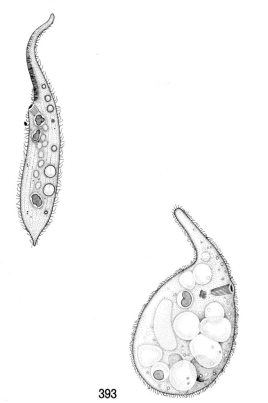

393

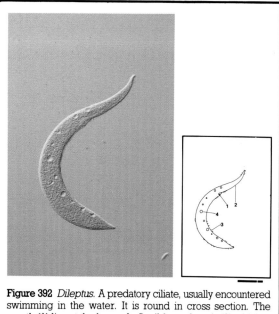

Figure 392 *Dileptus.* A predatory ciliate, usually encountered swimming in the water. It is round in cross section. The mouth (1) lies at the base of a flexible proboscis (2) which is swept through the water to increase the chances of contact with suitable prey. The proboscis contains numerous killing extrusomes, but these cannot be seen in this photograph. A few cilia (3) and some of the contractile vacuoles (4) are visible. *Dileptus* can be more than 1 mm in length, but this is unusual. *Differential interference contrast.* (Scale bar 100 µm.)

B Fat, highly vacuolated body with the mouth located at the base of a short, tapering neck. Body 150–400 µm long.
Fig. 393 TRACHELIUS

A Ciliated cells with colour.   GO TO 199

B Non-ciliated cells.   GO TO 187

A Cells firmly attached to the substrate.   GO TO 194

B Cells floating freely, or able to move slowly over the substrate by gliding or rolling.
HELIOZOA   GO TO 188

The heliozoa all have spherical bodies from which stiff arms radiate, supported internally by skeletal structures. They are predators that exploit diffusion feeding, i.e. they rely on prey cells swimming into them. The long arms increase the chance of contact with prey. Small bodies (extrusomes) move along the arms, and are used to hold potential prey.

Heliozoa are reviewed by Rainer (1969), Siemensma

(1981), and Page and Siemensma (1991). They are polyphyletic, three types being encountered in freshwaters (actinophryids, Step 188; centrohelids, Step 189B; and desmothoracids, Step 195). Many species have scales or spicules lying on the outside of the body. Numerous new species are being described (Dürrschmidt, 1985; Nicholls, 1983a & b; Nicholls and Dürrschmidt, 1985; Croome, 1986, 1987). Some filose amoebae (Step 83) may be confused with heliozoa.

A Large cells (body diameter greater than 100 μm) with a peripheral layer of large vacuoles.

Figs 394 & 395 ACTINOSPHAERIUM

Actinosphaerium is the multinucleated relative of *Actinophrys* (Step 189). The nuclei lie under the layer of vacuoles, and some arms may terminate on them. Actinophryids form cysts (Fig. 399). After excystment, uninucleate cells, which may be mistaken for *Actinophrys*, emerge. Members of this genus have been called *Echinosphaerium*. For an introduction to the literature, see Smith and Patterson (1986).

394

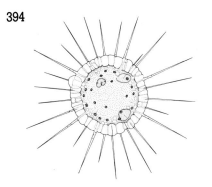

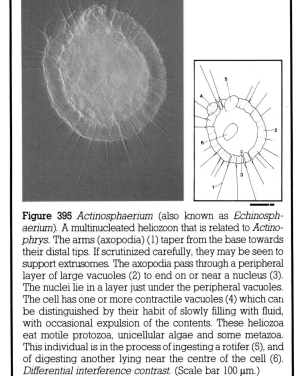

Figure 395 *Actinosphaerium* (also known as *Echinosphaerium*). A multinucleated heliozoon that is related to *Actinophrys*. The arms (axopodia) (1) taper from the base towards their distal tips. If scrutinized carefully, they may be seen to support extrusomes. The axopodia pass through a peripheral layer of large vacuoles (2) to end on or near a nucleus (3). The nuclei lie in a layer just under the peripheral vacuoles. The cell has one or more contractile vacuoles (4) which can be distinguished by their habit of slowly filling with fluid, with occasional expulsion of the contents. These heliozoa eat motile protozoa, unicellular algae and some metazoa. This individual is in the process of ingesting a rotifer (5), and of digesting another lying near the centre of the cell (6). *Differential interference contrast.* (Scale bar 100 μm.)

B Cell body less than 100 μm in diameter.

GO TO 189

A Tapering arms. Extrusomes are indistinct, and a central nucleus may be seen, especially in squashed cells. Body 30–90 μm.

Figs 396–399 **ACTINOPHRYS**

See Patterson (1979) for a general account of the biology of this genus, and Patterson and Hausmann (1981) for a discussion of feeding behaviour.

396

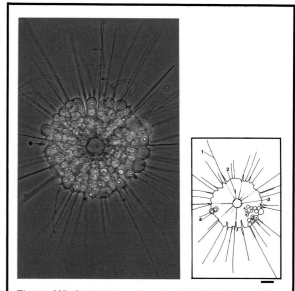

Figure 397 *Actinophrys*. A uninucleated heliozoon. The arms (axopodia) (1) taper from the base towards their distal tips. Small extrusomes, which are involved in the capture of food, may be seen moving on the arms and over the rest of the body. The microtubular axonemes, which support the arms, extend through the body, becoming visible (2) just before they terminate on the surface of the nucleus (3). The nucleus has peripheral nucleoli which make it rather obvious. This species eats flagellates and protozoa; these are digested within food vacuoles (4) which lie near the surface. The degree of vacuolation of the cytoplasm (here highly vacuolated) depends on the recent feeding history of the individual. *Phase contrast.*

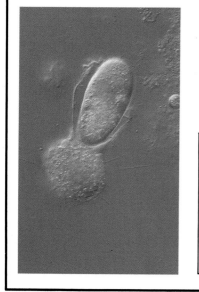

Figure 398 *Actinophrys* feeding. Small flagellates and ciliates constitute the usual diet of this heliozoon. The arms (2) often appear rough because extrusomes move along the skeletal support, deforming the overlying membrane. Extrusomes can sometimes be seen on the surface of the cell body (1). They are extruded as potential prey bumps into the arms; the prey (in this case, *Colpidium* (3)) is then held near the body. Next, a funnel-shaped pseudopodium emerges to envelop the prey. The leading edge (4) creeps over the cilia of the prey until it is enclosed within a food vacuole. Prey death and lysis occur only after complete enclosure. Many individual heliozoa may fuse together during feeding. *Differential interference contrast.*

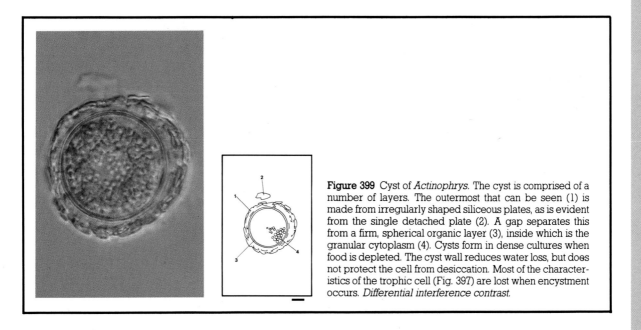

Figure 399 Cyst of *Actinophrys*. The cyst is comprised of a number of layers. The outermost that can be seen (1) is made from irregularly shaped siliceous plates, as is evident from the single detached plate (2). A gap separates this from a firm, spherical organic layer (3), inside which is the granular cytoplasm (4). Cysts form in dense cultures when food is depleted. The cyst wall reduces water loss, but does not protect the cell from desiccation. Most of the characteristics of the trophic cell (Fig. 397) are lost when encystment occurs. *Differential interference contrast.*

B Parallel-sided, delicate arms with prominent extrusomes.

THE CENTROHELID HELIOZOA   GO TO 190

In the centrohelid heliozoa the supports for the arms terminate on a centroplast, a small body lying in the centre of the cell. Most species have scales and/or spicules. See Bardele (1977) for an account of fine structure. It is to this group that many new species are being added (see notes and references after Step 187).

A Body coated with a layer of scales and/or spicules.   GO TO 192   **190 (189)**

B Without scales or spicules.   GO TO 191

A Body naked. Usually small (less than 30 μm).   Figs 400(a) & (b) OXNERELLA   **191 (190)**

Members of *Oxnerella* are most easily confused with *Heterophrys*, a genus characterized by its organic spicules (Step 192).

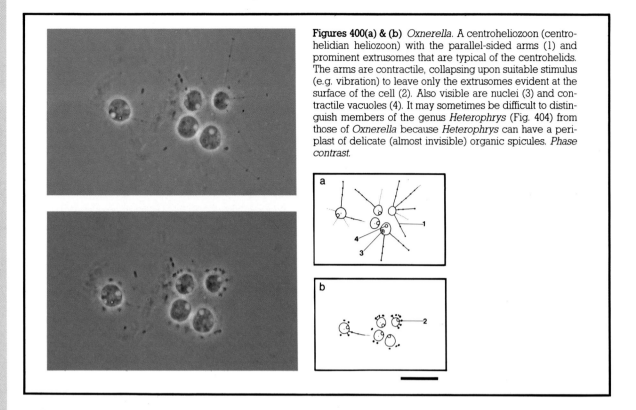

Figures 400(a) & (b) *Oxnerella*. A centroheliozoon (centrohelidian heliozoon) with the parallel-sided arms (1) and prominent extrusomes that are typical of the centrohelids. The arms are contractile, collapsing upon suitable stimulus (e.g. vibration) to leave only the extrusomes evident at the surface of the cell (2). Also visible are nuclei (3) and contractile vacuoles (4). It may sometimes be difficult to distinguish members of the genus *Heterophrys* (Fig. 404) from those of *Oxnerella* because *Heterophrys* can have a periplast of delicate (almost invisible) organic spicules. *Phase contrast.*

B With a layer of mucus around the cell. 12–30 μm in diameter.   Figs 401 & 402 **CHLAMYDASTER**

Members of *Chlamydaster* are very easily confused with some species of the filose amoeba *Nuclearia* (Step 83).

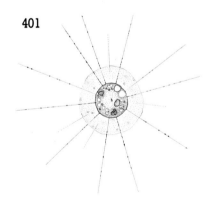

401

Figure 402 *Chlamydaster*. The distinguishing feature of this genus of centroheliozoa is the layer of mucus (1) that encases the cell. Through this layer pass the stiff arms (2) with their particle-like extrusomes. The genus is easily confused with some nucleariid filose amoebae, such as *Nuclearia delicatula* (Fig. 150). *Phase contrast.*

A Body (10–80 μm in diameter) coated with delicate inorganic spicules. Figs 403 & 404 HETEROPHRYS

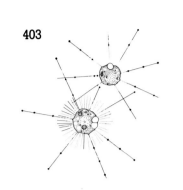

403

Figure 404 *Heterophrys*. This centroheliozoan genus is distinguished by the presence of organic spicules (1) around the body. Confusion with other centroheliozoan genera, such as *Acanthocystis* (Figs 411 & 412) or *Oxnerella* (Fig. 400), is possible. *Acanthocystis* can be distinguished by the presence of a layer of plate scales; and these are siliceous. Confusion with *Oxnerella* is possible because *Heterophrys* may have such delicate spicules that the cell may appear naked. The spicules of *Heterophrys* may only appear as an ill-defined halo around the cell. Some species are said to occasionally lose all their spicules. As with other species of heliozoa, the stiff arms (2) and extrusomes are used in the capture of food. *Phase contrast*.

B Body (10–150 μm in diameter) with scales and/or spicules. GO TO 193

A With flat scales only. Body 10–150 μm. Figs 405–407 RAPHIDIOPHRYS

See Siemensma and Roijackers (1988), and Patterson and Dürrschmidt (1988) for details of this genus.

405

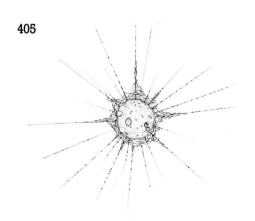

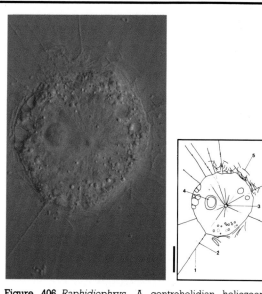

Figure 406 *Raphidiophrys*. A centrohelidian heliozoon (centroheliozoon). All heliozoa of this type have thin, parallel-sided arms (1) (axopodia) with prominent extrusomes (2). The microtubular axonemes that support the axopodia pass through the cytoplasm to terminate on a centroplast (3), a small body lying in the centre of the cell. The nucleus (4) is displaced to an eccentric position. Most of the organelles visible with the light microscope lie in the outer part of the cell. Several genera in this group are ensheathed by a layer (periplast) of siliceous spines and/or scales. This particular genus has scales only (5). *Differential interference contrast.*

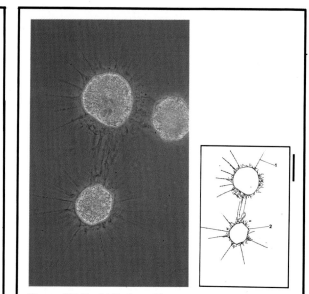

Figure 407 *Raphidiophrys*. The narrow arms with prominent extrusomes (1) are evident, as is the layer (periplast) of siliceous scales (2). This genus is distinguished by having flattened, siliceous scales, but no spicules. As with actinophryid heliozoa (e.g. Fig. 398), many cells may fuse together when feeding. After feeding, they draw apart and the scales (3) are distributed between the two cells. *Phase contrast.*

B With scales and spines.   GO TO 194

GO TO 194

**194
(193)**

A Spines are trumpet-shaped. Body 10–50 µm in diameter.   Figs 408 & 409 **RAPHIDOCYSTIS**

See Dürrschmidt and Patterson (1987), and Rainer (1968) for discussions of this genus.

408

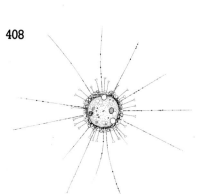

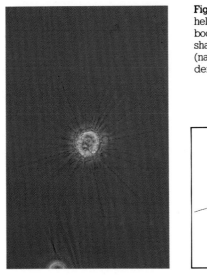

Figure 409 *Raphidocystis.* This genus of centrohelidian heliozoa has a layer of flattened plate scales lying close to the body surface, but it is distinguished by the radiating trumpet-shaped spine scales (1). The typical centroheliozoan arms (narrow, not tapering) and prominent extrusomes are evident (2). *Phase contrast.*

B Spines are forked or have a single point. Body 20–150 μm.

Figs 410–412 ACANTHOCYSTIS

See Nicholls (1983), Dürrschmidt (1985) Nicholls and Dürrschmidt (1985), Croome (1986, 1987) and Rees *et al.* (1980) for an introduction to the literature. More recently, some specialists have divided this genus on the basis of the ultrastructural features of the scales (Page and Siemensma, 1991).

410

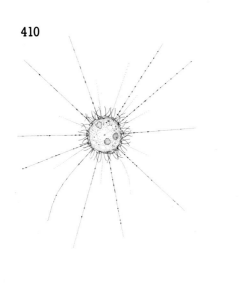

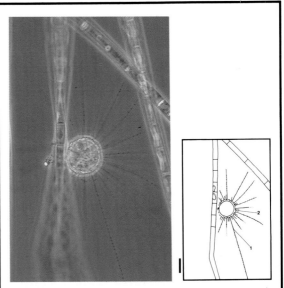

Figure 411 *Acanthocystis.* A centroheliozoon with the typical narrow centrohelidian axopodia (1) and prominent extrusomes. All species in this genus have siliceous scales and spines, which together make up the periplast (2). The different species are distinguished by the varying appearance of the spines and scales. This cell contains green chloroplasts, suggesting a recent meal of algae. Also visible are several filaments of the green alga, *Ulothrix. Phase contrast.*

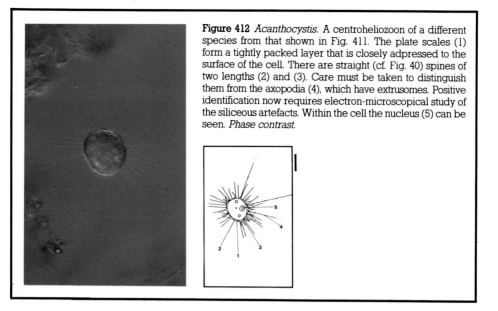

Figure 412 *Acanthocystis*. A centroheliozoon of a different species from that shown in Fig. 411. The plate scales (1) form a tightly packed layer that is closely adpressed to the surface of the cell. There are straight (cf. Fig. 40) spines of two lengths (2) and (3). Care must be taken to distinguish them from the axopodia (4), which have extrusomes. Positive identification now requires electron-microscopical study of the siliceous artefacts. Within the cell the nucleus (5) can be seen. *Phase contrast.*

195
(187)

A Amoeboid organism living in a stalked, perforated, organic test, through the pores of which extend fine stiff pseudopodia. The pseudopodia bear small granules or extrusomes. Lorica 25–100 μm in diameter.

Fig. 413 CLATHRULINA

Bardele (1972) describes the fine structure and general biology of this desmothoracid heliozoon. The life cycle has a flagellated and an amoeboid phase. Some aspects of the biology of a genus of smaller species (*Hedriocystis*; Fig. 414) are discussed by Bardele (1972) and Brugerolle (1985).

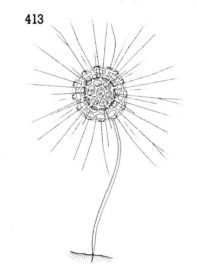

413

ALL SCALE BARS 20 μm UNLESS OTHERWISE INDICATED

Figure 414 *Hedriocystis*. A desmothoracid heliozoon. The amoeboid organism lives within a stalked (1) test (2). Delicate arms (3) bearing extrusomes extend through small apertures in the test. There may be a complex life cycle involving flagellated swarmers, an amoeboid stage, and an encysted phase. This genus contains relatively small species, but a second genus (*Clathrulina*) of much larger organisms may also be encountered. *Phase contrast.*

B Organisms with rounded or conical bodies from which radiate fine arms, each of which ends as a slightly swollen knob. The cells may be housed in a lorica. In some species the arms emerge in clusters, sometimes from raised regions of the body. SUCTORIA GO TO 196

Suctoria are ciliates that lack both locomotor cilia and feeding cilia. They have adopted a sessile life form, and most species prey on other ciliates. Each arm is an extended mouth: the knob at the end contains extrusomes that will hold food ciliates (Fig. 423). Once the food is captured, the cytoplasm is sucked down the arms into the body. A few suctoria have prehensile arms, used to collect debris. Ciliate affinities are evident from the fine structure and from the ciliated motile larvae (Fig. 424), which develop in a brood pouch, and/or bud from the apex of the cell. See Matthes *et al.* (1988) for a guide to the literature and genera.

A The body is attached to the substrate by means of a stalk. GO TO 197

196
(195)

B Without a stalk. Size variable, 10–400 μm. Fig. 415 TRICHOPHRYA

There are several genera of non-stalked suctoria. Others are, e.g., *Heliophrya* (Fig. 416) and the ectocommensal *Dendrocometes* (Fig. 417).

415

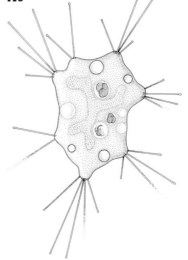

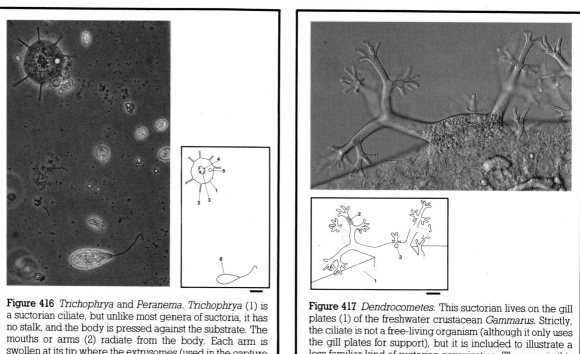

Figure 416 *Trichophrya* and *Peranema*. *Trichophrya* (1) is a suctorian ciliate, but unlike most genera of suctoria, it has no stalk, and the body is pressed against the substrate. The mouths or arms (2) radiate from the body. Each arm is swollen at its tip where the extrusomes (used in the capture of ciliates) are concentrated. Inside the cell the macronucleus (3), micronuclei (4) and contractile vacuole (5) can be seen. Also visible is a small species of the phagotrophic euglenid, *Peranema* (6) (see Fig. 71), with its single anterior flagellum. *Phase contrast.*

Figure 417 *Dendrocometes.* This suctorian lives on the gill plates (1) of the freshwater crustacean *Gammarus.* Strictly, the ciliate is not a free-living organism (although it only uses the gill plates for support), but it is included to illustrate a less familiar kind of suctorian organization. The arms in this genus are arborescent, i.e. they branch. In some arms the central canal (2), along which food passes, can be seen. Only part of the body of the ciliate, along with several contractile vacuoles (3), is in focus. *Differential interference contrast.*

A   The body is lodged in a test or lorica (20–300 μm long) shaped like a stalked egg cup.

Figs 418 & 419 **ACINETA**

See Bardele (1968, 1970) for detailed accounts, and Curds (1985a & b) for taxonomy.

418

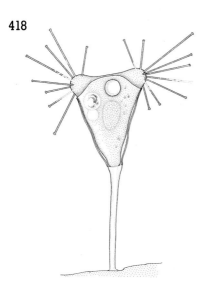

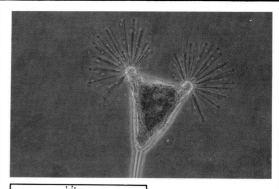

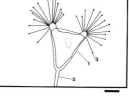

Figure 419 *Acineta.* A suctorian. Members of this genus are contained within an extracellular lorica (1) and supported on a stalk (2). The arms (mouths) (3) are very clearly arranged in clusters (fascicles). Each arm terminates in a broad knob which houses the extrusomes used to hold onto prey (other ciliates). *Phase contrast.* (Photo Tore Lindholm.)

B Without a lorica. GO TO 198

A The arms are grouped in clusters, and the body is drawn out where the clusters arise. Cells 20–200 μm long. Fig. 420 TOKOPHRYA

B The body is almost spherical, arms not clustered. Cells 15–250 μm long. Figs 421–424 DISCOPHRYA

420

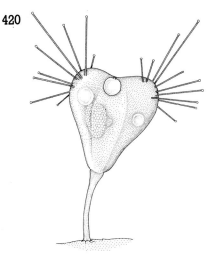

421

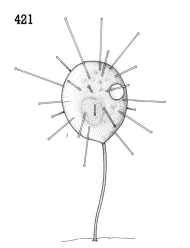

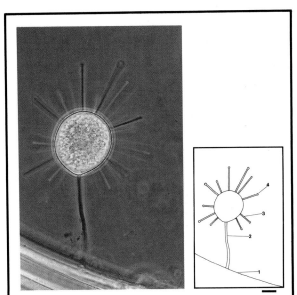

Figure 422 *Podophrya*. A suctorian ciliate. The trophic organism (as illustrated here) has no cilia. The organism is attached to the substrate (1) by an extracytoplasmic (secreted) stalk (2). A number of arms (3) radiate from the body, each of which is a mouth; the knobs (4) at the tips of the arms contain the numerous extrusomes used to hold onto ciliate prey. *Phase contrast.*

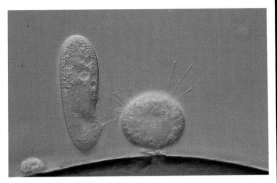

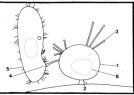

Figure 423 *Tokophrya* feeding. The individual shown here (1) is atypical because the stalk (2) is very short. The arms (mouths) (3) radiate in clusters from the aboral end. Each arm has a swollen tip where the extrusomes used in capturing ciliate food are located. One tentacle (4) has a firm grip on a ciliate prey, *Colpidium* (5). The ciliate remains alive as cytoplasm is sucked along the arm and into the body of the suctorian. As is usual after feeding, the cytoplasm of the suctorian looks granular. The prey may be released (alive) after feeding is complete. *Differential interference contrast.*

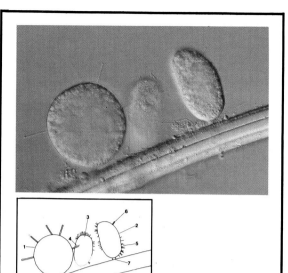

Figure 424 *Discophrya* and swarmer. Many sessile ciliates produce motile, ciliated larvae (also called swarmers) upon division. The swarmer, unlike the trophic cell, is able to seek out new sites in which to settle. This picture enables comparison of the trophic stage (1) with that of the swarmer (2). The trophic cell has no cilia and numerous arms, at least one of which (4) is attached to a ciliate (3), upon which the suctorian is feeding. After being released, the swarmer swims around until it finds a suitable place to settle. Metamorphosis then occurs, as shown here. Cilia (5) are still present, but the arms (6) and stalk (7) have begun to develop. *Differential interference contrast.*

199 A small number of ciliates are coloured. A green coloration is usually caused by symbiotic green algae. Most genera that have green species, also have colourless species (e.g. *Climacostomum*, Step 136; *Coleps*, Step 180; *Euplotes*, Step 138; *Frontonia*, Step 150; *Paramecium*, Step 172; *Prorodon*, Step 182; *Stentor*, Steps 118 and 163; *Vaginicola*, Step 130). Different species with symbiotic algae are often encountered together. Care should be taken to distinguish recently ingested algal food from symbionts: food particles will have different colours and be of different sizes, while symbiotic algae are similar in size and colour in any one cell. A list of ciliates with symbiotic algae is given in Christopher and Patterson (1984). Other colorations include pink (*Blepharisma*, Step 164; and some *Stentor* species, Steps 118 & 163) and orange (*Keronopsis*, a hypotrich, Step 136). There are also blue, brown and black species of *Stentor* (Steps 118 and 163), and some species of *Nassula* (Step 181) and *Loxodes* (Step 148) are golden. Certain algivorous ciliates, e.g. *Nassula*, may have a polka-dot pattern (Fig. 380).

Protozoan communities

Planktonic communities

The diversity of protozoa and the numbers of organisms in the water column of bodies of fresh water are usually a function of the amount of available organic matter. Oligotrophic lakes and over-wintering lakes typically have a sparse community of organisms, but this gets richer during the productive seasons and/or in richer eutrophic lakes. In enriched lakes, aggregates of bacteria and detritus may form in the water as a result of microbial activity; these aggregates may support a more diverse community which resembles that found in and on the benthos.

The water column typically contains autotrophic and heterotrophic flagellates. Colonial forms are common. In most cases the colonies are spherical (**1 & 8**); the feather-shaped colonies of *Dinobryon* (**2–4**) are atypical. The heterotrophic flagellates most usually consume bacteria, whereas the larger ciliates (**15, 16 & 19**) and heliozoa (**12 & 13**) may prey on flagellates. Most of the larger ciliates feed on small algae. If the oxygen content of the water is low, species of heliozoa, *Coleps* (**14**), *Euplotes* (**18**) and others with endosymbiotic green algae may be present. The algae produce photosynthates which may be used as food by their hosts, but the endosymbionts also generate oxygen which may secure the survival of their hosts.

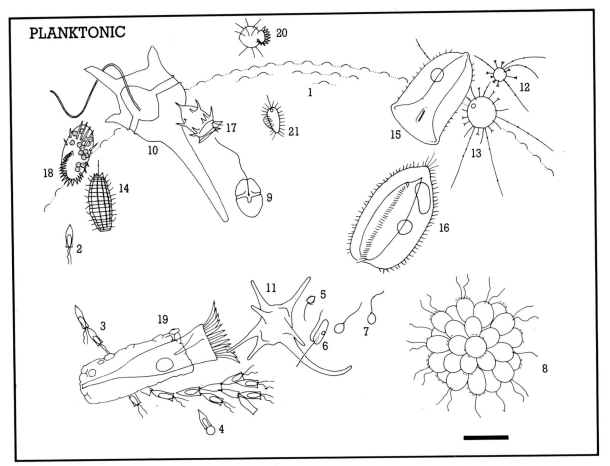

PLANKTONIC

1 *Volvox* (Figs 58–60), **2–4** *Dinobryon* (Figs 24 & 36), **5** *Paraphysomonas* (Figs 28, 100 & 101), **6** *Kathablepharis*, **7** *Trachelomonas* (Figs 116 & 117), **8** *Synura* (Figs 54 & 55), **9** *Gymnodinium* (Figs 137 & 138), **10** *Ceratium* (Figs 129 & 130), **11** *Amoeba radiosa* (Figs 141 & 142), **12** *Raphidocystis* (Figs 408 & 409), **13** *Acanthocystis* (Figs 410–412), **14** *Coleps* (Figs 375–377), **15** *Phascolodon* (Figs 314 & 315), **16** *Lembadion* (Figs 318 & 319), **17** *Hastatella* (Figs 372 & 373), **18** *Euplotes diadaleos* (Figs 259–261), **19** *Tintinnidium* (Fig. 370), **20** *Halteria* (Figs 304 & 305), **21** *Cyclidium* (Figs 331 & 332) (Scale bar 50 μm).

Attached communities

Submerged surfaces, whether inert, of plants or animals, or of detritus, often harbour rich and diverse communities of protozoa. Some species are permanently fixed to the surfaces; others browse over it.

Attached species typically remove food (in the form of suspended particles) from the surrounding water, by using flagella or cilia to create a flow of fluid and then extracting the bacteria with some kind of filter system. Heliozoa (**7 & 8**) adopt a different approach, relying on the movements of the prey, usually flagellates and small ciliates, to guarantee contact with the food. Food is trapped after it touches the 'adhesive' arms. Smaller species (flagellates, (**1–6**) and peritrich ciliates typically consume dispersed bacteria; the larger species prefer bigger food (usually other protozoa).

Many attached ciliates have specialized distributive stages (larvae or swarmers) which are produced after cell division. These stages are usually fully ciliated and will swim away from the parent cell, enabling the species to disperse and to colonize new habitats. Many larger species (**17–20**) have the ability to contract, a device which protects them from turbulence in the surrounding water, or from larger predators (flatworms and snails) which may glide over the submerged material.

Organisms that move over the immersed surfaces include hypostome ciliates, with their ventral mouths (**15**); hypotrich ciliates (**13 & 14**); and bodonid and euglenid flagellates (**7–10**). These usually consume individual attached- bacteria, other small adhering particles, diatoms, other algae, or filamentous cyanobacteria. 'Attached' protozoa may also be encountered in the water column, where they tend to be associated with aggregates of detritus.

ATTACHED

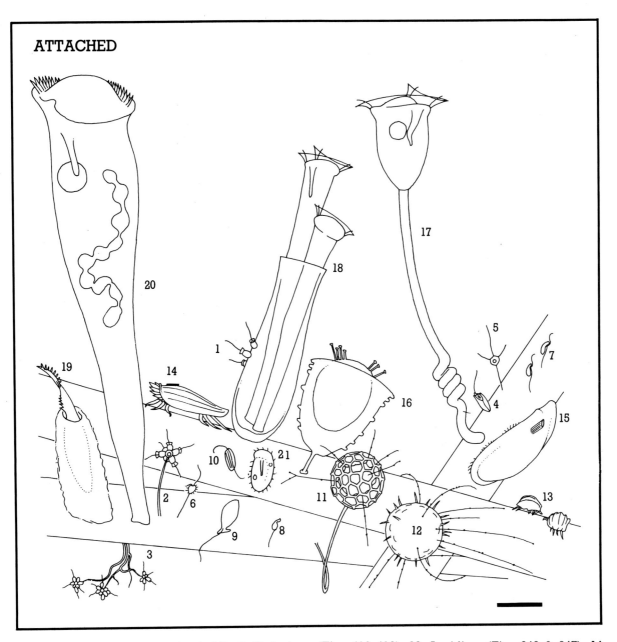

1 *Monosiga* (Figs 29(a) & (b) & 30), 2 *Codosiga* (Figs 43 & 44), 3 *Anthophysa* (Figs 45–47), 4 *Bicosoeca* (Figs 31 & 32), 5 *Actinomonas* (Figs 26 & 27), 6 *Paraphysomonas* (Figs 28, 100 & 101), 7 *Bodo* (Figs 25, 67–69), 8 *Rhynchomonas* (Figs 62 & 63), 9 *Urceolus* (Figs 87 & 88), 10 *Entosiphon* (Figs 75 & 76), 11 *Clathrulina* (Fig. 413), 12 *Acanthocystis* (Figs 410–412), 13 *Aspidisca* (Figs 246 & 247), 14 *Euplotes* (Figs 259–261), 15 *Trithigmostoma* (Fig. 299), 16 *Acineta* (Figs 418 & 419), 17 *Vorticella* (Figs 232–235), 18 *Vaginicola* (Figs 242 & 243), 19 *Stichotricha* (Figs 218 & 219), 20 *Stentor* (Figs 213–216), 21 *Chilodonella* (Figs 297 & 298) (Scale bar 50 μm).

Benthos

The benthic environment accumulates energy in two ways: from organic matter that settles from above; and by photosynthesis of organisms in the sediment. In most standing bodies of fresh water, the provision of energy through detritus usually exceeds that produced by photosynthesis. Consequently, there tend to be large numbers of bacteria and of the protozoa that feed on bacteria. The particular composition of the protozoan community depends on the season, the amount of organic matter, and the depth of the overlying water, etc.

The community illustrated here is of a type found in relatively clean waters. A wide range of taxonomic territory and of size is represented, from large amoebae (**12**) to flagellates that are no more than a few microns in size (**7**). Bacteria tend to be most numerous in the immediate vicinity of detritus, and the protozoa that exploit them are therefore usually found embedded in the detritus (**25**), loosely and temporarily attached to it (**1 & 5**), or gliding over it (the majority). Unlike the attached ciliates, the benthic community has relatively few species that are permanently fixed to the substrate. In contrast, their movement protects them from changing physicochemical conditions and allows them to escape submersion under fresh or disturbed detritus.

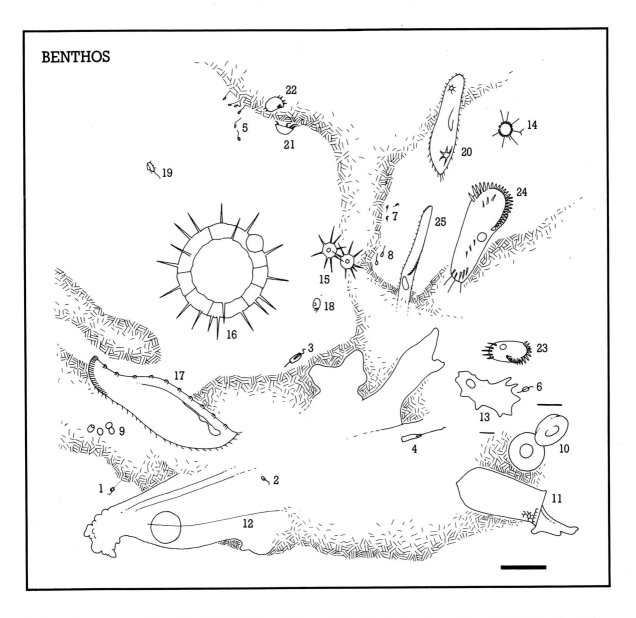

BENTHOS

1 *Paraphysomonas* (Figs 28, 100 & 101), 2 *Notosolenus* (Figs 79 & 80), 3 *Entosiphon* (Figs 75 & 76), 4 *Peranema* (Figs 70–72), 5 *Bodo* (Figs 25, 67–69), 6 *Protaspis*, 7 *Rhynchomonas* (Figs 62 & 63), 8 *Petalomonas* (Figs 82–84), 9 *Cryptodifflugia* (Figs 166 & 167), 10 *Arcella* (Figs 171–173), 11 *Difflugia* (Figs 186–188), 12 *Amoeba* (Figs 194–197), 13 *Mayorella* (Figs 192 & 193), 14 *Pompholyxophrys* (Figs 158 & 159), 15 *Actinophrys* (Figs 396–399), 16 *Actinosphaerium* (Figs 394 & 395), 17 *Loxophyllum* (Figs 282–284), 18 *Cinetochilum* (Figs 250 & 251), 19 *Cyclidium* (Figs 331 & 332), 20 *Paramecium caudatum*, 21, 22 *Aspidisca* (Figs 246 & 247), 23 *Euplotes* (Figs 259–261), 24 *Stylonychia* (Figs 256–258), 25 *Spirostomum* (Figs 321–323) (Scale bar 100 μm).

Organically rich benthos

Benthic habitats are periodically likely to accumulate large quantities of organic matter, such that the physiological demands made on the organisms become much greater; survival may depend on tolerance to low oxygen levels or acidity.

In organically overloaded sediments, bacteria and dissolved organic matter are the principal sources of food, and most species of indigenous protozoa are bacterivores or osmotrophs. Compared with well-aerated sites, there is little diversity in the community and familiar and widespread species become common. These species are largely the most weed-like of protozoa, in that they are the species which can usually be extracted from almost any permanent body of water. They appear in samples brought back to the laboratory, where higher temperatures stimulate the growth of bacteria and lower turbulence leads to a decline in oxygen levels. Often, these are the species (drawn from genera such as *Paramecium*, *Chilomonas*, and *Astasia*) which will grow readily in culture. Some of the species illustrated here are also encountered in waste-water treatment plants, which have a high content of organic matter.

1 *Helkesimastix* (Fig. 81), 2 *Bodo saltans* (Figs 25 & 69), 3 *Bodo caudatus*, 4 *Heteromita*, 5 *Cercomonas* (Figs 65 & 66), 6 *Chilomonas* (Figs 96 & 97), 7 *Trepomonas* (Figs 104(b) & 105), 8 *Hexamita* (Figs 103, 104(a) & 106), 9 *Bodo saltans* (Figs 25 & 69), 10 *Spumella* (Fig. 102), 11 *Polytoma* (Fig. 95), 12 *Astasia* (Figs 89 & 92), 13 *Cryptodifflugia* (Figs 166 & 167), 14 *Colpidium* (Figs 337–339), 15 *Glaucoma* (Figs 248 & 249), 16 *Paramecium caudatum*, 17 *Paramecium putrinum* (Fig. 348), 18 *Cyclidium* (Figs 331 & 332), 19 *Halteria* (Figs 304 & 305), 20 *Holosticha* (Figs 268 & 269), 21 *Diplophrys* (Figs 146 & 147), 22 *Spirostomum teres* (Figs 321 & 322), 23 *Loxodes* (Figs 280 & 281) (Scale bar 50 µm).

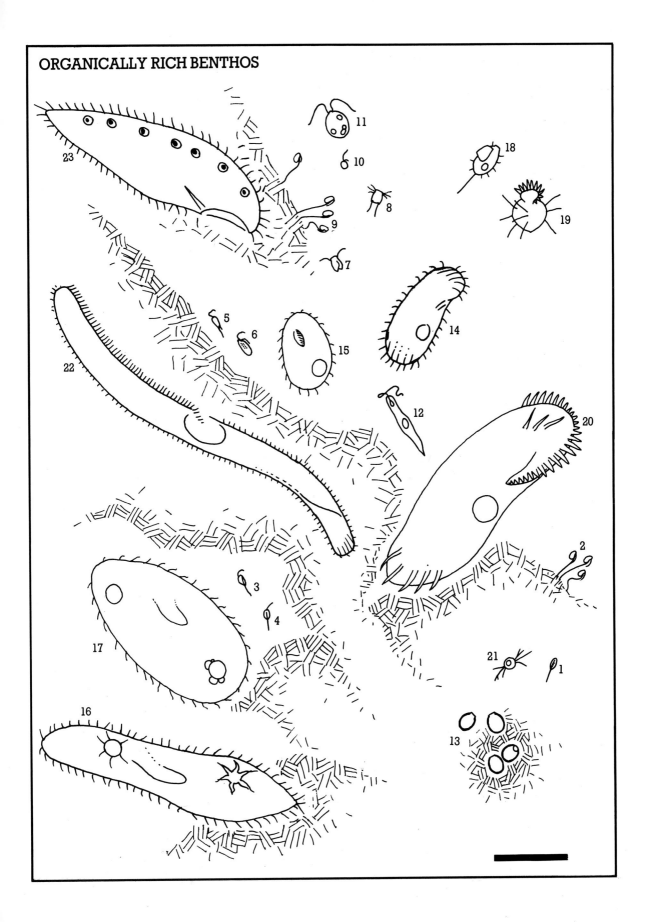

Anoxic benthos

As the organic loading in a habitat rises, so the amount of oxygen required by the microbial community, which degrades the organic matter, rises. In benthic environments, oxygen is only supplied from above. As it diffuses into the sediments, it is consumed by the microbial community. At a certain depth, no further oxygen is available for microbial respiration. At this point, which varies in position from above the sediment to many metres below the sediment surface, the habitat becomes anoxic.

Deeper in the sediment, the metabolism of the microbial community has to rely on other compounds to take over the role of oxygen as a terminal electron-acceptor. Ultimately, carbon dioxide and various sulphur compounds are used (with methane and sulphides as by-products); metal salts, especially iron, are converted to metal sulphides (usually black) and hydrogen sulphide gas (which smells bad to initiates, but promising to the cognoscenti) is given off. This kind of environment is referred to as being 'reduced'. The physiological conditions are very different to those of oxygenated areas, and most organisms that require oxygen die rapidly if placed in a reduced habitat.

Protozoologically, the reduced environment is very interesting as a variety of specialized protozoa exist in such conditions. Some of these (pelobionts and diplomonads, **1–6**) are believed to have evolved before oxygen was available on the earth, and to have survived in anoxic environments ever since. However, most, for example the ciliates (**7–14**), have adapted aerobic metabolism to suit life in reduced habitats, a major benefit of which is reduced competition for rich supplies of food.

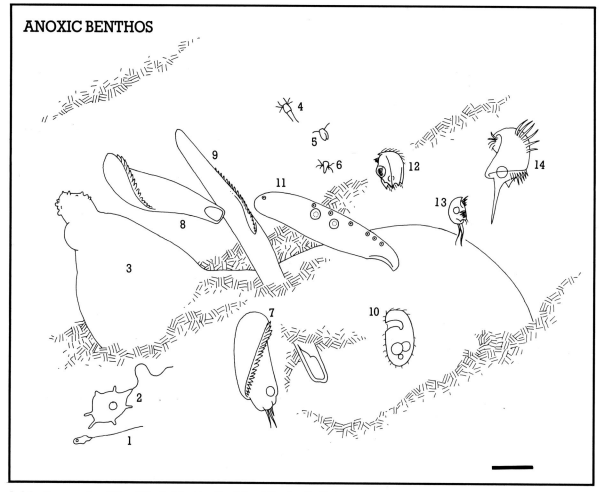

ANOXIC BENTHOS

1 *Mastigamoeba* (Fig. 81), **2** *Mastigella* (Fig. 86), **3** *Pelomyxa*, **4** *Hexamita* (Figs 103, 104(a) & 106), **5** *Trepomonas* (Figs 104(b) & 105), **6** *Trigonomonas*, **7** *Brachonella* (Fig. 310), **8** *Metopus* (Figs 308 & 309), **9** *Spirostomum* (Figs 321–323), **10** *Plagiopyla*, **11** *Loxodes* (Figs 280 & 281), **12** *Saprodinium*, **13** *Myelostoma*, **14** *Caenomorpha* (Figs 306 & 307) (Scale bar 50 µm).

Sewage treatment plants

Sewage treatment plants receive water containing dissolved and particulate organic matter and bacteria in suspension. The purpose of the plant is to remove this matter, a process usually carried out by a microbial community of (principally) bacteria and protozoa. The community removes the incoming organic matter by converting it to a form (flocs, slime, etc.) that can easily be separated from the fluid, with the result that a relatively clean effluent is produced.

Because this process is biological, it is sensitive to factors which may affect organisms. The process will change as a function of the temperature, the nature of incoming material, pollutants, and the nature of the microbial community, etc. Two factors have a major influence on the biological community and the performance of the process: the length of time that incoming sewage is exposed to microbial processing (the retention or residence time); and the concentration of organic matter being added to the process (organic loading).

In normal water bodies, microbial communities that receive an input of organic matter are normally subject to a predictable sequence of change, a 'succession', which begins with the growth of bacteria. Predators that eat bacteria, grow fast, and are most tolerant of the consequences of high levels of organic matter (low oxygen, acidity) will appear next – usually this means small flagellates. The flagellates will reduce the numbers of bacteria so that the demand for oxygen is lessened and the water body can become more oxygenated. This reduces some of the physiological constraints on organisms, allowing the community to become more diverse. Rapidly growing ciliates which also eat bacteria, and some small amoeba are usually next to appear. They are followed by slow-growing ciliates and amoebae, which are often specialized to eat restricted types of food (e.g. filamentous bacteria, other ciliates, etc.). Given that environments are not homogeneous, it may be possible to find a few representatives of any part of the succession at any one time.

Most sewage treatment works are biological systems which are subject to an input of organic matter. Because the input is continuous, the normal development of a succession does not occur. Instead, the succession is terminated at a stage that is determined by the rate of flux of fluid through the system. The passage of fluid through the system removes some of the organisms of the community and any species which cannot reproduce quickly enough to compensate for such a loss will be removed from the system. Thus, sewage treatment systems with fast flow rates will tend to favour or-

ganisms with rapid rates of reproduction, usually the smaller protozoa such as flagellates (Zone A) or small ciliates (Zone B). If the passage of fluid is slow, then a greater diversity of organisms is likely to appear, ultimately extending to metazoa which, in contrast with protozoa, have slow rates of reproduction (Zones D and E).

An increase in the organic loading leads to a higher demand for oxygen. Oxygen levels in the process may become depleted, and the plant may even become anoxic despite mechanisms to keep it aerated. Over-loaded systems will tend to harbour organisms that prefer anoxic conditions (Zone A) (pelobionts, 1; diplomonads, 2). Heavily loaded systems with low levels of free oxygen will favour those flagellates, amoebae, and small ciliates (3–17) normally found in organically polluted habitats. As the loading of organic matter declines, so more normal conditions prevail, and the diversity of organisms which may live under these conditions increases. The number of individuals encountered is usually smaller in underloaded plants.

The zones illustrated over the page indicate the kinds of organisms that are likely to be prevalent under particular circumstances. The boundaries between zones/communities are not clear cut. The organisms in Zone A are mostly anaerobes or microaerophiles and are found in high-rate sewage treatment works that take in very high concentrations of organic matter. Such plants are usually found in urban areas, where the pressures of population tend to necessitate a rapid flux of material through the plant. Such plants often produce a turbid (= poor) effluent.

Zone B contains many bacterivorous species which are tolerant of organic pollution; these organisms are usually found in plants with a moderately high organic loading and with a fair quality of effluent in that the organic matter separates readily from the fluid. Sewage plants with communities dominated by organisms in Zone C and above usually produce a relatively high quality of effluent. The types and diversity of organisms indicate a relatively long residence time and, therefore, the opportunity for the community to metabolize and convert incoming organic matter. Systems harbouring organisms mostly in Zones D and E must have a fairly long residence time (e.g. ditch systems). Such systems produce minimum amounts of sludge and so reduce transport and disposal costs. However, they are inefficient in that relatively small amounts of sewage are treated, and at a low rate, and they can either be given more organic matter or be run at a faster rate.

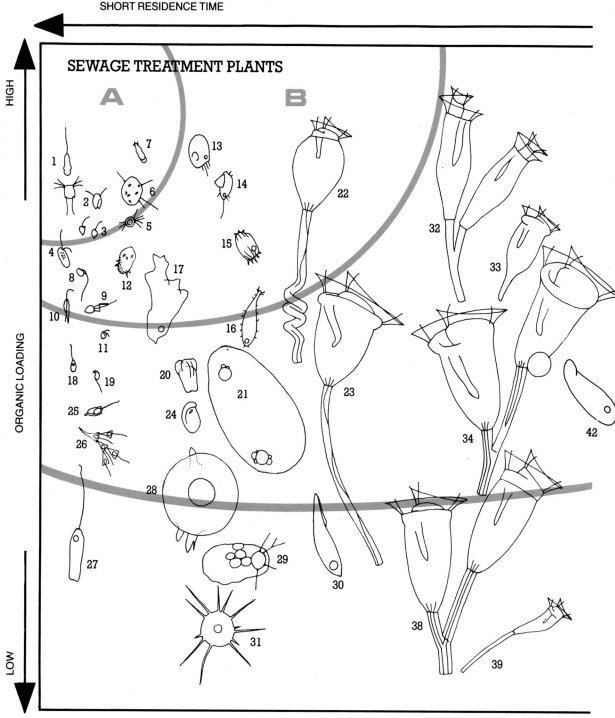

HIGH

ORGANIC LOADING

LOW

SEWAGE TREATMENT PLANTS

A   B

1 *Mastigamoeba* (Fig. 81), 2 *Hexamita* (Figs 103, 104(a) & 105) and *Trepomonas* (Figs 104(b) & 105), 3 *Paraphysomonas* (Figs 28, 100 & 101) and *Oicomonas*, 4 *Chilomonas* (Figs 96 & 97), 5 *Diplophrys* (Figs 146 & 147), 6 *Pamphagus* (*Chlamydophrys*), (Figs 168 & 169), 7 *Vahlkampfia*, 8 *Bodo saltans* (Fig. 69), 9 *Monosiga* (Figs 29(a) & (b) & 30), 10 *Cercomonas* (Figs 65 & 66), 11 *Goniomonas* (Figs 98 & 99), 12 *Cochliopodium* (Figs 161 & 162), 13 *Cinetochilum* (Figs 250 & 251), 14 *Cyclidium* (Figs 331 & 332), 15 *Aspidisca* (Figs 246 & 247), 16 *Trachelophyllum*, 17 *Mayorella* (Figs 192 & 193), 18 *Petalomonas* (Figs 82–84), 19 *Rhynchomonas* (Figs 62 & 63), 20 *Thecamoeba* (Figs 209 & 210), 21 *Paramecium putrinum* (Fig. 348), 22 *Vorticella microstoma*, 23 *Vorticella* spp. (Figs 232–235), 24 *Chilodonella* (Fig. 298), 25 *Bicosoeca* (Figs 31 & 32), 26 *Poteriodendron*, 27 *Peranema* (Figs 70–72), 28 *Arcella*

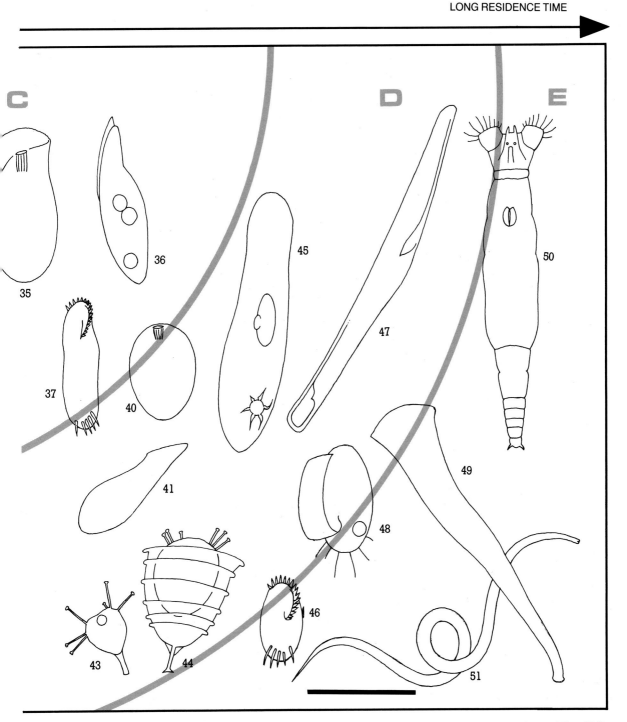

(Figs 171–173), **29** *Trinema* (Figs 180 & 181), **30** *Litonotus* (Figs 285–287), **31** *Actinophrys* (Figs 396–399), **32** *Opercularia* (Figs 226 & 227), **33** *Rhabdostyla* (Fig. 237), **34** *Carchesium* (Figs 223–225), **35** *Trithigmostoma* (Fig. 299), **36** *Hemiophrys*, **37** *Oxytricha* (Fig. 263), **38** *Zoothamnium* (Figs 220–222), **39** *Vorticella hamata*, **40** *Prorodon* (Figs 384–386), **41** *Spathidium* (Figs 275 & 276), **42** *Colpidium* (Figs 337–339), **43** *Tokophrya* (Fig. 420), **44** *Acineta* (Figs 418 & 419), **45** *Paramecium caudatum* (Fig. 334), **46** *Euplotes* (Figs 259–261), **47** *Spirostomum* (Figs 321–232), **48** *Pleuronema* (Figs 333 & 334), **49** *Stentor* (Figs 213–216), **50** rotifer (Fig. 15), **51** nematode (Fig. 18) (Scale bar 100 µm).

191

Glossary of terms

Aboral: relating to position – away from the mouth (cf. **Adoral**).

Acronematic: relating to flagellum – with thinner region near the tip (distal).

Actinophryid: a type of heliozoon (see Step 188) with tapering arms, e.g. *Actinophrys* and *Actinosphaerium*.

Actinopod: pseudopodium with an internal support, usually of microtubules; or organisms (heliozoa, radiolaria) with such pseudopodia.

Activated sludge plant: part of one type of modern sewage treatment process in which sewage is violently aerated, and which (typically) supports large populations of ciliated protozoa.

Adaptive strategy: a suite of linked properties of an organism, which together make it more competitive. Often said of evolutionary trends in which convergence is evident (e.g. sessile habit, colonial habit, etc.).

Adhering: attached.

Adoral: relating to position – near the mouth (cf. **Aboral**).

Adoral zone of membranelles: a band of membranelles found in polyhymenophoran (spirotrich) ciliates (Step 116). It extends from the front of the cell to the cytostome. Mostly used for feeding, but occasionally for locomotion.

Agar: a commercially available gelling material used as a basis for cultures of bacteria, fungi and some protozoa. Add 1.5 g per 100 ml of water. Heat indirectly in a water bath. When molten, pour into Petri-dishes or equivalent containers.

Agglutinated: sticking/stuck together. Often said of the tests of some protozoa, which are made from pieces of debris (xenosomes) that adhere to each other (Fig. 187).

Algae: organisms capable of photosynthesis without relying on symbiotic organisms. May be prokaryotic = bacterial (blue-green algae = cyanobacteria), or eukaryotic, in which case mostly (some say exclusively) protists. Said of, e.g. green algae, brown algae, diatoms, chrysophytes (chrysoflagellates) and cryptophytes (cryptoflagellates).

Aloricate: no lorica present.

Amoeba: protists that move using pseudopodia.

Amoeboid: like an amoeba. Usually means that the organism has the capacity to produce pseudopodia.

Amphizoic/amphitrophic: said of organisms capable of gaining energy and nutrients by both autotrophic and heterotrophic means (= mixotrophic).

Ampullae: part of the contractile vacuole complex of some protists. Distensible channels found in the vicinity of contractile vacuoles (Fig. 288).

Anastomose: said of pseudopodia that branch and fuse, thereby forming a network. As in *Biomyxa* (Fig. 144).

Animals: multicellular eukaryotes, mostly with cells arranged in epithelia (layers attached to collagenous base), gaining energy and nutrients by ingesting particles of food; or clearly related to such organisms. The invertebrates and vertebrates. Inappropriately used in reference to protists, e.g. when protozoa are referred to as unicellular animals.

Anoxic: no free oxygen present, a situation that commonly arises in natural habitats when the biological demand for oxygen exceeds the supply, e.g. in sediments or when a site is organically polluted.

Anterior: the front part of the cell. Usually determined as being in the direction of normal movement.

Aperture: an opening. Used in relation to tests or loricae to refer to the site of emergence of pseudopodia, flagella, or cells.

Apex: the most anterior point of a cell.

Apical: pertaining to the apex (the anterior pole).

Arborescent: relating to type of colony with a tree-like, branching pattern.

Arm: long, thin, non-motile projection from a cell, pseudopodia of heliozoa (Fig. 397) or feeding tentacles of suctoria (Fig. 419).

Athalamida: a type of amoeba with filose pseudopodia and no shell. Usually regarded as being related to foraminifera.

Attached: adhering to a fixed point of the substrate, either permanently (cannot easily detach) or temporarily (can easily let go).

Atypical: unusual. Normally said of an organism that is quite unlike other members of its group.

Autotrophic: organisms that trap energy from physical or chemical sources and use the energy to assemble the macromolecules of which they are made, e.g. photosynthesis (cf. **Heterotrophic**).

Axenic: said of cultures, to indicate that no other species is/are present, that is, without bacteria, pure.

Axis: a conceptual line passing from the anterior pole of the cell to the posterior pole.

Axoneme: a geometrically packed assemblage of microtubules (subcellular scaffolding) used to support flagella, the arms of heliozoa, etc.

Axopodia: pseudopodia with internal skeletal structures made of microtubules.

AZM: abbreviation for adoral zone of membranelles.

Back: dorsal (cf. **Ventral**).

Bacteria: prokaryotic organisms. Typically used to refer to those not having photosynthetic pigments, but strictly including the blue-green algae (cyanobacteria).

Bacterivore: eats bacteria (= bactivorous).

Bacterivorous: said of bacterivores.

Basket: a cylindrical assemblage of microtubular rods, which surrounds the mouth of some cyrtophore ciliates (Fig. 379), and is used during the uptake of food.

Benthic: associated with the benthos.

Benthos: the bottom sediments of rivers, lakes, ponds, etc..

Bicosoecida: a type of filter-feeding flagellate (Fig. 32).

Biodisc: part of a process of sewage treatment. Usually supports a film rich in protozoa.

Biogenic: produced by living organisms (e.g. biogenically derived macromolecules).

Black mud: reduced muds found below the surface of sediments in lakes, rivers, etc. The blackness is caused by the occurrence of metal sulphides.

Bloom: dense growths of organisms, usually algae. Typically short-lived and typically of one species.

Blue-green: a colour due to the photosynthetic pigments of prokaryotic algae. The bluish tinge is caused by accessory photosynthetic pigments (Fig. 4).

Blue-green algae: the only kind of alga that is prokaryotic. Some prefer the terms 'cyanobacteria' or 'blue-green bacteria' to emphasize prokaryotic affinities.

Bodonid: a type of small flagellate (Fig. 68) related to trypanosomes.

Bright-field optics: a method of setting up microscopes to gain maximum resolution. Contrast is low, so suitable for stained material, not suitable for observing most protozoa (Fig. 349).

Buccal: relating to mouth structures (e.g. buccal ciliature).

Capitate: with a head. Said of tentacles of, e.g. *Mesodinium* (Fig. 274).

Cell: mass of cytoplasm bounded by plasma membrane. Of two types: prokaryotic or eukaryotic. Eukaryotic organisms are the animals, plants, fungi and protists. Most protists are comprised of a single cell, and with protists, the term 'organism' is synonymous with 'cell'.

Cellulose: polysaccharide used to make the walls (normally around the outside) of certain types of cell.

Centric: a type of diatom (Fig. 8) exhibiting radial symmetry, e.g. *Stephanodiscus* and *Melosira*.

Centrohelid: a type of heliozoon, with thin, parallel-sided, arms and with prominent extrusomes, e.g. *Acanthocystis* (Fig. 411).

Chlorophyll: a family of pigments used in photosynthesis to trap radiant energy. Normally located within chloroplasts. Chloroplasts with chlorophyll b have a bright green colour, while those with chlorophylls a and c are off-green or yellow-green. See **Colour**.

Chloroplast: an organelle found in eukaryotic algae and plants (and occasionally as symbionts in certain protozoan and animal cells). The site of photosynthesis and of chlorophyll. See **Colour** and **Chlorophyll**.

Chonotrich: a type of ciliate found as an ectosymbiont on crustacea (Fig. 381).

Chromatic aberration: a fault in microscopes that leads to a failure in the object being imaged faithfully. Colours are poorly imaged so that the 'wrong' colour is seen, or a colour may be seen when none is present in the object.

Chromosomes: assemblages of the molecule DNA in nuclei of eukaryotic cells. Genetic information is located on chromosomes. Chromosomes of dinoflagellates have a peculiar arrangement, and can be observed in living cells (Fig. 132).

Chrysomonad: a type of flagellate (Fig. 23): autotrophic, mixotrophic or heterotrophic. Plastids if present with chlorophylls a and c, with two unequal flagella. Also referred to as chrysos or chrysophytes.

Chytrids: a type of protist believed to be related to fungi. With a flagellated stage which settles on other organisms, sending a branching system of 'rhizoids' into the host cell, and using them to draw up energy and nutrients. Flagellated stage has a single long posterior flagellum.

Cilium: a behavioural type of eukaryotic flagellum, distinctive because it occurs in large numbers, has a co-ordinated behaviour, and usually directs fluids parallel to the surface.

Cingulum: a horizontal or spiral groove on the surface of dinoflagellate cells, in which lies a flagellum.

Circumferential: relating to the circumference. Usually means passing round the cell in a plane normal to the (longitudinal) axis of the cell (as of **Cingulum**).

Cirrus: a locomotor structure typical of hypotrich ciliates, formed from a tight cluster of individual cilia (Fig. 261) that move as a single entity.

Coccoid: rounded in shape, ball-like.

Collar: a thin flange encircling a structure, e.g. the collar of pseudopodia around the flagellum of collar flagellates (Fig. 29).

Collar flagellate: a type of flagellate (Fig. 29), e.g. *Monosiga*.

Collecting canal: part of the contractile vacuole complex of certain ciliates. One or more of these structures may lead from the cytoplasm to the contractile vacuole (Fig. 345).

Colonial: a type of organization in which many cells are bound together by secretions or cytoplasmic extensions. The cells are usually similar, and so compete with each other for resources, but some degree of differentiation (cellular specialization) does occur in a number of colonial species. Colonies are usually spherical if planktonic, or arborescent if attached.

Colour: colour in protists can be a useful diagnostic feature. It may be caused by: photosynthetic pigments in chloroplasts (e.g. Fig. 121); by other pigments in the cytoplasm (Fig. 325); by metal salts which accumulate in secretions (e.g. *Trachelomonas*, Fig. 117); or may be artificially created by chromatic aberration. Many colour-blind people will not be able to distinguish between different pigment combinations in chloroplasts.

Colpodid: a type of ciliated protozoan (Fig. 328), e.g. *Colpoda*.

Compound microscope: a type of microscope with a selection objective and eyepiece lenses, as illustrated on p. 15.

Condenser: part of a compound microscope situated between the light source and the specimen. Used to focus light onto the specimen.

Conjugation: a type of sexual event during which two cells fuse (Fig. 344). It may or may not lead to reproduction.

Contract: a kind of cellular motility in which the whole or part of a cell shortens at a visible rate in one or more directions.

Contractile vacuole: part of the contractile vacuole complex; often the only part that is visible with the light microscope. Collects fluid and periodically allows the fluid to be discharged through the cell surface (Fig. 283).

Contractile vacuole complex: an organelle involved in osmoregulation in protist cells. Comprises a contractile vacuole, spongiome (a membranous system not usually visible with the light microscope), possibly a pore, collecting canals, and ampullae.

Convergence: refers to similarity in form (or other features of organisms) that has been achieved independently, usually as a means of adapting the organism to a particular life style, e.g. the star-shape of actinophryids (Fig. 397) and centrohelids (Fig. 411).

Crawling: a type of movement in which the organism moves across the substrate while maintaining continuous contact with it. May involve no visible organelles (gliding), or cilia or flagella.

Crenulated: with a regularly indented margin.

Crustacea: a type of arthropod (metazoan), including copepods and ostracods.

Cryptomonad: a type of flagellate (Fig. 126) with two similar flagella emerging from an anterior depression, and with extrusible ejectisomes associated with the depression. Also referred to as cryptos and cryptophytes.

Cyanobacteria: blue-green algae.

Cyrtophorine: a type of kinetofragminophoran ciliate with a nasse associated with the mouth (Fig. 298), e.g. *Chilodonella*.

Cyst: a differentiated state in which the body is enclosed within a continuous extracellular lorica, and exhibits very little activity (Figs. 152 & 399). Exploited only by some species. Often used to increase chances of survival in unfavourable conditions.

Cytoplasm: the matter that makes up cells, within which organelles occur.

Cytopharynx: part of the food ingestion apparatus (mouth) of some cells. Usually a channel of microtubules that draws newly formed food vacuoles away from the cytostome and into the cell (Fig. 342).

Cytoproct: the site at which old food vacuoles fuse with the cell surface, and undigested residues are excreted. Found in some ciliates.

Cytoskeleton: intracellular components used to give shape to a cell, or to create tracts along which cellular organelles may be moved. Mostly comprised of microtubules and actin filaments.

Cytostome: literally, 'the cell mouth'. Used only in reference to organisms that ingest food at one or more particular locations (and then best used in reference to the region(s) of the cell surface through which food gains entry into the cell). Part of the 'mouth' structures. See also **Cytopharynx**.

Dark ground: a type of imaging in microscopy, in which the object appears bright against a dark background (Fig. 353).

Daughter colonies/daughter cells: the products of the cell division of protists.

Desiccation: drying out. Dehydration.

Dichotomous: a pattern of branching in colonial organisms in which one element (stalk) gives rise to two equal and divergent branches. Also used in reference to identification keys in which the identity of an organism is established by presenting questions for which there are only two acceptable answers.

Differentiation: the act of becoming specialized (differentiated) in form or function. Protists may be specialized to feed (trophonts), to weather unfavourable conditions (cysts), or to hunt out new resources (theronts). Each of these states is achieved through differentiation.

Diffusion feeding: feeding strategy in which the predator relies on the movements of the prey to make contact, as in heliozoa and suctoria.

Distal: away from (cf. **Proximal**).

Division: the most common mechanism of reproduction of protists, in which a cell replicates itself by dividing into two. The plane of division of ciliates is usually across the cell body (transverse, Fig. 341); flagellates usually divide longitudinally (Fig. 92).

Desmid: a type of green alga (Figs 9 & 10).

Desmothoracid: a sessile protist in which a heliozoan-like organism is located within a lorica (e.g. *Clathrulina*, Fig. 413).

Detritovore: eats detritus.

Detritus: fragments of dead plant and animal material before, during and after breakdown by agents of decay. May incorporate inorganic matter (such as mud).

Diagnostic: used in relation to a particular characteristic (feature) of an organism, which is quite distinctive and therefore can be used to identify that organism.

Diatom: a kind of protist with chloroplasts and a siliceous lorica/wall (Fig. 6). Of two kinds: centric and pennate. Common and widespread.

Differential interference contrast: a type of imaging used in light microscopy, in which the boundaries of refractive index difference are revealed as a light–dark boundary (Fig. 352). Ideal for the study of protists, as a very thin optical slice is taken through the specimen so that organelles are shown clearly (also called Nomarski).

Diplomonad: a type of flagellate (Fig. 104) with two nuclei and two sets of four flagella, e.g. *Trepomonas* and *Hexamita*.

Dispersed: said of bacteria that float freely or swim in a fluid environment (in contrast to **Attached** or **Adhering**).

Dissecting microscope: a microscope with relatively low magnifying powers, but with a very wide field of view and a long working distance between the object and the objective lenses. Suitable for low-power scanning of samples. Also referred to as a binocular microscope.

Dorsal: refers to the back of the cell, i.e. the face of the cell away from the ventral (cf. ventral). The concept does not always apply.

Ectosymbiotic: an organism living on the surface of another organism.

Ejectisome: a type of explosive extrusome found in cryptomonads.

Elongate: relatively long shape (length more than three times greater than the breadth).

Emergent: referring to emerging or coming out, e.g. euglenids may have two flagella, but only one may project from the front of the cell, i.e. the emergent flagellum.

Encyst: to change from a trophic or other state to a cyst (cf. **Differentiation**).

Envelope: used to refer to enclosures, such as a mucilaginous sheath enclosing a cell body. Also said of the space used by the flagellum when beating normally. In profile, the beat envelope may have a peculiar shape which may assist in the identification of an organism.

Equal: said of flagella when they are similar in length and in behaviour (cf. **Unequal**).

Equatorial groove: a groove that passes around the middle of the cell in a plane at right angles to the longitudinal axis of the cell.

Eruptive: refers to a type of movement by lobose pseudopodia, in which the pseudopodia progress by a sequence of sudden bulges, rather than by a gradual and continuous flow.

Euglenid: a type of flagellate (Step 32), mostly with an ability to squirm, and/or with a helically sculpted body. Some have chloroplasts. Typically with rather thick flagella, e.g. *Euglena* and *Peranema*. Also referred to as Euglenophyceae.

Euglenoid motion: a kind of squirming motion typical of some euglenids. Also called metaboly.

Eukaryotic: refers to cells with nuclei and other membranous organelles, the plants, animals, fungi and protists.

Eumycetozoa: a type of slime mould (Fig. 20).

Excrete: to eject materials from cells. Best used in reference to undigested residues of food from food vacuoles, or in reference to fluid (cf. **Secrete** and **Extrude**).

Excyst: to change from an encysted state to another viable state (cf. **Differentiation**).

Exoskeleton: a supportive structure lying outside the cell or body.

Extracellular: outside the cell.

Extrude: to push out.

Extrusome: a kind of organelle, the contents of which can be extruded, e.g. to catch or kill prey, or for protection.

Extrusopodia: pseudopodia that bear extrusomes (Fig. 404).

Eyepiece: the lens(es) of a microscope that cast(s) an image into the eye. Also referred to as oculars.

Eyespot: a structure found in some flagellated algal cells. Usually a red or orange collection of oil droplets. Believed to be able to detect light and to influence the movement of the cell. Also referred to as a stigma.

Facultative: optional, said of a state which an organism may or may not adopt, depending on circumstances (cf. **Obligate**).

Field of view: the area that may be seen when looking down the eyepieces of a microscope. The width of the area depends on the design of the microscope and the choice of objective. However, it is always the same for a given objective on one microscope and so can be used to make rough estimates of cell size.

Filament: a thin strand. May refer to the appearance of an organism, a strand of cytoplasm, or to the thin metal wire in a bulb, which is heated to emit light.

Filose: refers to a type of thin, thread-like pseudopodium without an internal skeleton (Fig. 151).

Filter feeding: a type of feeding in which suspended particles are consumed. Requires a propeller (usually cilia or flagella) to direct a current of water to the cell, and a filter (or other) device to concentrate the particles before enclosure within a food vacuole.

Fix: to attach (e.g. to the substrate), or to kill and preserve.

Flagellum: a filamentous structure used for motion. Flagella of prokaryotes are biochemically, structurally and functionally very different from flagella of eukaryotes. The eukaryotic flagellum has a skeletal component comprised of microtubules, and is flexed by an interaction of the protein dynein with the microtubules. The microtubules form a cylindrical structure, the axoneme, inside the flagellum. Cilia are a modified kind of eukaryotic flagellum. Prokaryotes have a stiff flagellum that is rotated to propel the cell.

Flagellar membrane: the membrane enclosing the axonemal part of a eukaryotic flagellum.

Flagellar pocket: a depression in the cell surface of euglenids and cryptomonads, at the base of which are inserted the flagella. May be referred to as a reservoir.

Flagellate: a kind of protist that bears flagella. Refers to a very diverse group with unclear boundaries. Distinguished from ciliates because the flagella are few in number, and usually create a thrust along the length of the organelle, rather than parallel to the body surface.

Flatworm: a type of metazoan (Fig. 14) that uses cilia and muscles for movement, and ingests food through a muscular channel, the pharynx. Soft bodied.

Flotation: pertaining to floating. Protists may be modified to favour floating rather than sinking. Typical adaptations to flotation include long arms or spines.

Focal plane: a plane in which the image produced by a microscope is in focus. Used, e.g., in reference to the location of the shutter of a camera.

Food web: an ecological concept conveying a sense of the complex interactions within a community, where organisms can indirectly or directly affect others by acting as food or as predators, or by competing with each other for food.

Free-living: referring to organisms that live in inert habitats (water bodies, soils, sands, etc.), move, and gain their food without relying on the intervention of other species (as opposed to parasitic or symbiotic). The category does not have sharp boundaries.

Free-swimming: refers to an organism that is able to move freely through the fluid phase of medium. Compare with crawling or gliding, where the organism requires contact with the substrate.

Front: the anterior. That part of the cell projecting forwards when the cell is moving.

Frustule: the siliceous lorica of a diatom.

Gastrotrich: a type of metazoan. Resembles some ciliates (Fig. 16).

Gelatinous: with a jelly-like consistency.

Genus: a taxonomic rank above the level of species (a genus contains one or more species). Organisms carry two names (e.g. *Pompholyxophrys punicea* or *Paramecium aurelia*), usually written in italics. The first name is the genus name, and the second is that of the species. The genus name may be abbreviated to a single letter where the meaning is unambiguous (*P. aurelia* or *H. sapiens*). Strictly, all genera have different names, but occasionally the same name is used twice (e.g. for a plant and for an animal).

Gliding: a type of movement in which the organism remains in contact with the substrate. The movement is gradual, and the contribution made by individual organelles of locomotion is not evident (cf. **Crawling**).

Golden: a colour usually used to describe chloroplasts that contain chlorophylls a and c (as in chrysomonads). Other colours are green, off-green or blue-green.

Granules: solid inclusions in cells, or items adhering to the surface of cells. Usually refractile, in that they may look bright when viewed with the microscope.

Granuloreticulose: refers to a type of pseudopodium usually found in the marine foraminifera. The pseudopodia have a granular texture and form a network (see *Lieberkuehnia*, Fig. 170), e.g. granuloreticulose amoebae.

Graticule: a glass disc with a scale, grid or other pattern etched into the surface. Placed in the eyepiece of a microscope, it is calibrated against a scale on a slide and used (mostly) to measure the sizes of microscopic objects.

Green: a colour used to describe chloroplasts, indicating that chlorophyll b is present. Plastids without chlorophyll b are less bright in colour, and may be off-green, golden, or blue-green. Chloroplasts that contain chlorophyll b are found in euglenids, green algae (including volvocids) and some symbiotic algae of ciliates and amoebae.

Green algae: the Chlorophyceae – algae with green chloroplasts and an external cell wall made of cellulose. This group includes desmids, volvocids and filamentous green algae, and gave rise to the land plants.

Gromiida: a type of amoebae with filose pseudopodia and a lorica.

Groove: a long, shallow depression on a surface.

Gullet: a term that is sometimes used to refer to a depression on a cell surface, particularly one into which flagella are inserted (see **Flagellar pocket**). Not a desirable term as it misleadingly suggests a role in food uptake.

Haptorid: a type of ciliate that preys on other protozoa, usually other ciliates, capturing and/or killing them with explosive extrusomes (e.g. *Didinium*, Fig. 263).

Helical: of or like a helix. Like a spiral, but extending in the third dimension.

Helioflagellates: flagellates with stiff, radiating arms around the flagellum. Of two categories: pedinellid (e.g. *Actinomonas*, Fig. 27) and dimorphid.

Heliozoon: a type of protist with stiff, radiating arms (e.g. *Acanthocystis* and *Actinophrys*, Figs 412 & 397). The heliozoan body form has evolved on several occasions.

Heterocyst: a differentiated, thick-walled cell, found in the filaments of some blue-green algae.

Heterotrich: a type of polyhymenophoran ciliate (Step 116) that moves by means of individual cilia arranged in kineties.

Heterotrophic: refers to a mode of nutrition in which the consumer relies upon molecules created by other organisms for energy and nutrients. Of two categories: osmotrophic (absorbing soluble organic matter) and phagotrophic (ingesting particles of food).

Hymenostome: a type of oligohymenophoran ciliate with buccal ciliature comprised of three membranelles and an undulating membrane (e.g. *Colpidium*, Fig. 338).

Hypostome: a type of ciliate with the mouth located ventrally. Used in reference to cyrtophorine ciliates that use a basket of nematodesmata to aid the ingestion of food (e.g. *Pseudomicrothorax* and *Trithigmostoma*, Figs 292 & 299).

Hypotrich: a type of polyhymenophoran ciliate (Step 136) that moves using cirri.

Idiosome: a structure produced by the organism, as opposed to a xenosome or foreign body. Used to refer to the elements that comprise (or adhere to) the tests of some amoebae.

Immersion oil: a special kind of oil used in microscopy, a drop of which is used between the coverslip and front face of some objectives.

Immotile: not moving. Sometimes used to refer to cells that are fixed to the substrate.

Inclusions: structures located within cells, such as food vacuoles, granules, etc.

Ingesta: items that have been ingested by a protist, usually while they retain their integrity so that they may be identified.

Ingestion apparatus: the equipment used by cells to aid in the ingestion of food. Normally used in relation to rods of microtubules lying near the cytostome (mouth), but may also include extrusomes and the cytopharynx.

Ingestion organelle: ingestion apparatus.

Ingestion rods: individual components of the ingestion apparatus, as in cyrtophorine (hypostome) ciliates (Fig. 298) and some euglenid flagellates (*Peranema*, Fig. 71).

Inorganic: not organic (carbohydrate, protein, etc.). Refers to, e.g. sand, mud, silica, etc.

Interference contrast: a type of imaging for light microscopy, in which areas of differing optical properties appear as different colours (Fig. 354).

Intracellular: inside the cell.

Iris: part of a microscope. A diaphragm that may be closed or opened at will. Usually there is one iris in the lamp housing and one in the condenser.

Jump: a type of movement exhibited by some cells, characterized by sudden (instantaneous) changes in position (e.g. *Halteria*, Fig. 305).

Karyorelict: a type of ciliate in which the macronuclei are unable to divide (e.g. *Loxodes*, Fig. 280).

Kick: a type of movement exhibited by cells that sit in one position, but may jerk or flick part of the cell body (e.g. *Bodo saltans*, Fig. 69).

Köhler: a microscopist whose name is associated with a type of illumination that guarantees maximum brightness (but low contrast); it involves centering all optical components on the optical axis of the microscope (the axis of the objective), and positioning the condenser to focus the light onto the specimen.

Lamp housing: part of a microscope containing the lamp. Now built into the body of compound microscopes, it was previously a free-standing structure (and still is for some dissecting microscopes). Older microscopes used mirrors to direct the light into the microscope from independent lamps, candles, or the sun.

Large: a description of size, the exact meaning of which depends on the range encountered in the group of organisms under consideration. Bodonids are relatively small organisms; thus a cell with one dimension measuring 20 µm would be thought large. In contrast, among the ciliates, a cell measuring 20 µm would be regarded as small, and a ciliate would need to be well over 100 µm before being thought of as large. Usually, protists with one dimension that is 50–100 µm or more are referred to as large.

Lens tissue: a special kind of tissue recommended for cleaning glass surfaces in microscopes and other optical instruments. Available from opticians and chemists, as well as from laboratory suppliers. Each sheet should be used once and then discarded.

Lip: a shallow ledge or a long protrusion from a cell.

Lobose: a type of pseudopodium that is relatively broad (Fig. 139).

Longitudinal: refers to the axis of the cell from the front to the posterior of the cell.

Lorica: an organic or inorganic casing or shell, incompletely surrounding an organism. Usually loose fitting. Sometimes called a test.

Macronucleus: one of two types of nuclei found in ciliates. Typically, the larger of the two. It may be rounded, either like a long sausage or like a string of beads. It is involved in the production of proteins, but not in sexual reproduction. Essential for the day-to-day activities of the ciliate (cf. **Micronucleus**).

Magnification: a measure of the enlargement of an image relative to an object. Normally, it is simpler and more informative to refer to the real size of the object.

Marginal row: relating to the cirri of hypotrichs (Figs 254 & 260), referring to the one or two rows nearest to the lateral margin of the cell. May be continuous around the posterior end of the cell (one row), or broken at the posterior end (two rows).

Mastax: part of the digestive system of rotifers (Fig. 15). A grinding structure lying just behind the mouth.

Mastigamoebae: a kind of flagellate with a single flagellum and a cell body that produces pseudopodia. Usually from anoxic sites (e.g. *Mastigamoeba*, Fig. 85).

Matrix: relating to the structure or consistency of material.

Measuring eyepiece: a type of microscope eyepiece that includes an etched scale. After calibration against a micrometer slide, the eyepiece may be used to measure the size of microscopic objects.

Median: near the centre (as of cell, e.g. median nucleus).

Medium: the fluid environment in which protists live, or the solution of salts and other materials in which they are cultured. Also used to describe a relative size of organism, dependent upon the group under consideration. Medium-sized flagellates are 10–30 μm, and medium-sized amoebae and ciliates are 40–100 μm.

Medusoid: umbrella-shaped – like a jelly-fish.

Meiosis: a form of nuclear division usually associated with sexual activity.

Membranelle: a compound structure comprised of many cilia, and associated with the mouth of a ciliate. Either present in groups of three (Oligohymenophora, Fig. 342) or as a band of many more (Polyhymenophora, Fig. 256).

Metaboly: change. Used either in reference to change in molecules (metabolism), or to change in cell shape, as in euglenoid motion.

Metal salts: soluble or insoluble chemical compounds that incorporate one or more metal ions, such as iron, manganese, calcium or sodium.

Metazoa: animals (cf. **Protozoa**).

Microaerophilic: preferring low levels of dissolved oxygen.

Micrometer slide: a glass slide with a scale (usually 1 mm) etched on its surface. Used to calibrate the field of view or the scale in a measuring eyepiece, such that the size of objects viewed with a microscope can be measured.

Micrometre: a unit of length equivalent to one millionth of a metre. Abbreviated to μm.

Micron: a micrometre.

Micronucleus: one of two kinds of nuclei found in ciliates, dividing to produce two similar nuclei during asexual reproduction, and producing nuclei with half the complement of DNA for sexual activity. Usually the smaller of the two types of nuclei, but many may be present. Some cells lack a micronucleus, surviving quite well without it (cf. **Macronucleus**).

Microscope: in the context of this Guide, a device with glass or plastic lenses, the function of which is to produce a magnified image of an object. Of two common types: compound and dissecting. In a broader context, the above are kinds of light microscope. However, microscopes producing images from sound, beams of electrons, etc. are also available.

Microtubule: a subcellular structure comprised of the protein tubulin. Used for support, it is part of the cytoskeleton. Individual microtubules cannot be seen by conventional light microscopy, but aggregates of microtubules can.

Mixotrophic: used in reference to organisms that use a mixture of nutritional strategies, e.g. organisms that have chloroplasts and carry out photosynthesis, but which are also able to feed by phagocytosis.

Mobiline: a type of peritrich ciliate without a stalk attaching it to the substrate (Fig. 373).

Morphology: the shape and form of an organism or part of an organism.

Motile: moving, e.g. by swimming, gliding, crawling, jumping or kicking. Part of the body (e.g. cilia) may be motile in a cell which is not motile and which is fixed in one position.

Mouth: part of the body involved in the acquisition and internalization of food. The mouth usually includes a cytostome, but may also involve elements external (e.g. buccal cilia) or internal (e.g. ingestion rods, extrusomes, cytopharynx) to the cytostome.

Mucilaginous: made of, or with the texture of, mucus.

Mucus: a jelly-like substance produced by organisms. Texture may vary from being virtually fluid to stiff and rubber-like.

Mycelia: the organization of the feeding stage of fungi, in which the cytoplasm is enclosed within a radiating system of walled tubes or hyphae.

Naked: used in relation to cells that have no cell wall, lorica, or other coating.

Nasse: cylindrical ingestion apparatus of some ciliates. The wall is made up of rods of microtubules. Also referred to as a basket.

Neck: a narrow, anterior part of the body, often with an ingestion apparatus at the anterior end. Also used to refer to a narrow region of a lorica or test, leading to the opening(s).

Nematodesmata: stiff aggregates of many microtubules, found around the cytostome of some ciliates (e.g. Fig. 298) and used during the ingestion of food. A type of ingestion rod.

Nematode: a type of metazoon. Typically smooth-bodied (Fig. 18).

Neuston: the environment of the interface between water and air. It is often rich in bacteria and protists.

Neustonic: associated with the neuston.

Nomarski: a Polish (later French) microscopist who gave his name to a type of contrast enhancement (see **Differential interference contrast**).

Nomenclature: the terminology of a science. A system of names for objects. In biology, nomenclature usually refers to the rules governing the names of animals and plants, embodied in the International Codes of Zoological and Botanical Nomenclature.

Non-motile: not moving. May be said of a whole cell which may yet have motile parts, or of organelles.

Nucleolus: an optically dense region (or regions) in a nucleus, associated with RNA synthesis. Not always visible.

Nucleus: an organelle found only in eukaryotic cells, in which most of the cellular DNA (genetic material) is located. Most cells have a single nucleus, but certain species may have many.

Nudipodia: a type of unsupported pseudopodium without evident extrusomes (cf. **Extrusopodia**).

Nutrient: that which provides nutrition. The nutrients of heterotrophic organisms are primarily biogenically derived macromolecules, whereas those of autotrophs are usually more simple compounds, such as dissolved phosphates and nitrates.

Objective: a magnifying lens of a microscope. Most microscopes carry a selection of objectives. The objective is located near the object being viewed.

Obligate: said of a state which an organism must adopt (cf. **Facultative**).

Ocular: refers to the lens of a microscope, into which one looks when trying to view the image of an object. Also called an eyepiece.

Odontostome: a type of polyhymenophoran (spirotrich) ciliate that is distinguished by a flattened, sculpted body and few somatic cilia (e.g. *Epalxella*, Fig. 312).

Off-green: a colour used to describe some chloroplasts that lack chlorophyll b, but which usually have an olive or brownish hue.

Oligohymenophora: a type of ciliate with buccal ciliature comprised of an undulating membrane and three membranelles, e.g. peritrichs (Step 119), *Colpidium* (Figs 337–339), etc.

Oligotrich: a kind of polyhymenophoran ciliate (Step 178) in which the adoral zone of membranelles is used not only in the acquisition of food but also to propel the cell (e.g. *Halteria*, and tintinnids, Figs 305 & 370).

Oomycete: a type of fungus. The mature organism takes the form of walled mycelia, but produces flagellated stages in the life cycle. Related to the chrysomonads.

Opisthe: posterior daughter cell produced by transverse division of a ciliate.

Organelle: a discrete structure found within eukaryotic cells.

Organic: chemical compounds that have been produced by organisms. Said mostly of sugars and polysaccharides, proteins, etc., it essentially refers to the hydrocarbons. Does not include the inert chemical compounds, such as silica, produced by some organisms.

Osmotrophic: refers to a form of nutrition in which soluble compounds are taken up by the organism, either by pinocytosis or by mechanisms capable of transporting one or a few molecules at a time (membrane pumps).

Overloaded: used in reference to sewage works that have more incoming sewage than the plant is designed for (cf. **Underloaded**). Overloaded systems are equivalent to polluted natural environments, and usually have a poor quality effluent.

Oxidize: To add oxygen.

Parasitism: an association of organisms in which one partner benefits to the detriment of the other.

Particles: small items (e.g. bacteria, protists, or inert material such as clay). May be in suspension in fluid, be associated with detritus, or be used in the manufacture of loricae/tests. Referred to as 'particulates' by some people.

Pellicle: the outer region of cytoplasm of some protozoa. The term is applied only when the region can be distinguished because it appears to be relatively stiff and highly structured. The term 'cortex' may also be used. Used mostly in reference to euglenids and ciliates.

Peniculine: a type of oligohymenophoran ciliate that is distinguished by the particular arrangement of the membranelles (e.g. *Paramecium*, Fig. 345).

Pennate: refers to one of two types of diatom (Fig. 5). Without radial symmetry, and often able to move by gliding.

Periplast: the entire assemblage of scales, spines and spicules that encases some heliozoa, chrysomonads, etc.

Peristalsis: regular contractions of a body or part of a body. Used mostly in the context of the intestinal system of vertebrates, but also refers to the squirming behaviour of some euglenids.

Peristome: the region of the body around, and external to, the mouth. Strictly, the region must be modified to favour the acquisition of food.

Peritrich: a type of oligohymenophoran ciliate (Step 19) in which one membranelle and the undulating membrane are greatly lengthened, spiralling around the oral end of the body (e.g. *Vorticella*, Fig. 233).

Petri dish: a low, flat, circular dish with vertical sides. Made of glass or plastic, it is used extensively for the culture of micro-organisms, and is similar in shape to many centric diatoms.

Phagocytose: to take food by phagocytosis, i.e. to ingest visible particles of food by enclosing them with a membrane to form a food vacuole.

Phagotroph: an organism that feeds by phagocytosis.

Pharynx: a region of the ingestion apparatus that lies internal to the mouth of a metazoan organism, or internal to the cytostome of a protist. Involved in the swallowing process (see **Cytopharynx**).

Phase contrast: a method of contrast enhancement used widely in light microscopy. It is particularly useful in protozoology. In this process, areas with differing refractive indices appear darker or lighter than the background (Fig. 351).

Photomicrography: the process of taking photographs through a microscope.

Photosynthates: the products of photosynthesis.

Photosynthesis: a means of acquiring energy for metabolism. It involves trapping radiant energy in chloroplasts, the use of that energy to break up water molecules (hydrolysis), and the conversion of released energy into an accessible form, such as the molecule ATP. The only form of autotrophy in eukaryotic cells. Some heterotrophic protists have symbiotic algae that allow them to exploit photosynthesis.

Photosynthetic pigments: large molecules in chloroplasts. They absorb radiant energy, hence they have colour. Mostly chlorophylls and carotenes (and, occasionally, phycobilins).

Phycobilin: a type of photosynthetic pigment, mostly found in blue-green algae, red algae and some cryptomonads.

Pigments: Molecules that appear coloured.

Pinocystosis a process of ingesting material by enclosing it with a membrane. The resulting structure is usually too small to be seen with the light microscope, and is mostly suitable for the ingestion of fluid or mucus.

Pipette: a glass or plastic tube designed to facilitate the process of dispensing liquid. May be graduated, so that measured volumes can be dispensed. Pasteur pipettes have a narrow tip, but can be 'pulled' to a finer tip after heating, so that they can be used for handling small volumes of liquid or individual cells. Dropping pipettes are available from chemists and pharmacists, and have a rounded tip for safety.

Planar: in one plane, e.g. said of flagellar beating.

Plankton: organisms living in the water column (above the sediment).

Planktonic: from the plankton. If protists, they may either be swimming or floating.

Plant: multicellular eukaryote with the capacity for photosynthesis using chlorophyll b, and with cells surrounded by cellulosic walls. As with the term 'animal', 'plant' is sometimes inappropriately applied to unicellular organisms. To state that algae are unicellular plants is confusing; it would be more accurate to say that algae are unicellular organisms with some specified characteristics (e.g. photosynthesis) that are also found in plants (or that plants are multicellular organisms with some specified characteristics that are also found in algae).

Plasmodium: a type of amoeboid organization involving a large mass of cytoplasm and, usually, many nuclei. A type of body form adopted by some slime moulds.

Plastid: another word for 'chloroplast'. From it come the terms 'aplastidic' and 'plastidic', meaning with and without chloroplasts respectively. See notes after Step 5.

Platyhelminth: a flatworm.

Podite: a narrow extension from the posterior end of some ciliates (e.g. Fig. 295), used for adhesion. Foot-like.

Polarizing microscopy: a type of microscopy that produces bright images of objects with a crystalline substructure. The process relies on the use of polarized light, hence the term.

Polyhymenophora: a type of ciliate (Steps 136 & 161) in which the buccal ciliature includes more than three membranelles. Also called a spirotrich.

Polysaccharide: a fairly large molecule comprised of many sugar molecules (chemically) joined together. Polysaccharides are relatively inert and may be used by cells to form external walls, etc. Polysaccharides include cellulose and starch.

Pomiform: shaped like an apple (as of some helioflagellates).

Posterior: that part of the body away from the direction of normal movement, or away from the mouth. The term 'back' may also be used, but there is a possibility of confusion with 'dorsal'.

Paraxial rod: a rod of material lying within the flagellum parallel to the axoneme. It is only found in some protists (e.g. euglenids), and causes the flagellum to appear relatively thick (Fig. 71).

Prokaryotic: refers to a type of organism, the cells of which are without nuclei or other membrane-bound organelles, i.e. bacteria.

Prostome: a type of ciliate with the mouth located at the anterior end of the body. Usually ingests larger particles of food (e.g. *Coleps*, Step 376).

Proter: anterior daughter cell produced by transverse division of a ciliate.

Protostelid: a type of slime mould. Spores usually produced by a single cell rather than by a mass of cytoplasm, as is the case for most slime moulds.

Protozoa: those heterotrophic and (a few) autotrophic protists that have, by tradition, been studied by protozoologists (see p. 9). Some people prefer the broader term 'protist', but 'protozoa' is still widely used to refer to flagellates, ciliates, amoebae, and sporozoa without chloroplasts.

Proximal: near to (cf. **Distal**).

Pseudopodium: a transient extension of the cell surface, used for locomotion or feeding. Pseudopodia may be supported internally (actinopods) or not (rhizopoda), they may be thread-like (filose) or broad (lobose), may or may not bear extrusomes (extrusopodia and nudipodia respectively), and there may be one (monopodial) or many (polypodial) produced at one time.

Punctate: with a dimpled or spotted appearance.

Pusule: a system involving a sac and channels, found in some dinoflagellates. The function is not understood, but it may act as an osmoregulatory organelle.

Pyrenoid: a protein body lying inside some types of chloroplasts.

Raphe: a slit in the siliceous shell of most pennate diatoms. Motile diatoms always have a raphe; it appears to be involved in the secretion of mucus that pushes the cell around.

Raptorial: refers to a type of feeding in which the consumer moves around in search of suitable nourishment (cf. **Diffusion** and **Suspension feeding**).

Recurrent flagellum: a flagellum that curves posteriorly from its site of insertion, to trail along and/or behind the body.

Reduced: a chemical state of some environments, in which free oxygen is absent and if made available will be chemically consumed by the molecules present. Typically, a site rich in sulphides (e.g. hydrogen sulphide) and methane, usually black in colour and with a strong smell. Also applied to some parts of cells to indicate small size or absence (reduced flagella are short).

Refractile: capable of refracting light, thereby acting like an irregular lens. Refractile granules may appear bright when viewed with a microscope, but they may also be a source of chromatic aberration, appearing coloured when they are not.

Reservoir: a part of euglenid cells. A depression from which the flagella arise and into which the contractile vacuole empties its contents. Also referred to as a flagellar pocket.

Rhizopoda: those amoebae that have pseudopodia with no stiffening axoneme (Step 72). Polyphyletic.

Ridge: Slightly raised, elongate region of a structure.

Rotifer: a type of metazoon, also called a 'wheel animal' because of the two clusters of motile cilia around the mouth (Fig. 15).

Sapropelic: said of sites (or of organisms inhabiting such sites) that are very rich in organic matter and that (usually) lack oxygen. Usually refers to sediments.

Scale: a measuring device (as on a micrometer slide), or a flat, plate-like structure produced by some protists.

Sculpting: used to refer to a body that has a fixed, irregular shape, e.g. grooved, ridged or spiny.

Scuticociliate: a type of oligohymenophoran ciliate (Step 168). Typically with a well-developed, undulating membrane (e.g. *Cyclidium*, Fig. 332).

Secrete: to expel an artefact or material fabricated in a cell to the exterior, e.g. scales, spines, mucus. Compare with **Excrete**, which refers to the extrusion of bodies originally produced elsewhere and temporarily ingested.

Sessile: refers to organisms that are fixed to the substrate, by e.g. a stalk or lorica.

Sheath: a loose covering (usually of mucus). A more rigid or substantial covering would be referred to as a wall or lorica.

Siliceous: of, relating to, or incorporating silica.

Sine wave: contiguous waves, e.g. in a beating flagellum, that are equally spaced and all of the same height (amplitude).

Size: see **Small**, **Medium** and **Large**.

Skip: a type of motion that is characteristic of some flagellates. The cell moves near the substrate, progressing quickly for a short distance, slowing down as the flagellum touches the substrate, and then speeding up as the contact between flagellum and substrate is lost. May transform into swimming or gliding.

Slime mould: a type of amoeboid organism that produces a walled stage (cyst or spores) at the end of an elevated stalk. The life cycle may involve flagellated stages, or organisms resembling conventional amoebae.

Small: a relative size of organism; the absolute size depends on the group of organisms under consideration. A flagellate of 5 µm or less is considered to be small, but ciliates up to 20 µm would also be described as small.

Solitary: organisms occurring individually, not in colonies.

Somatic: relating to the body surface, e.g. somatic ciliature (cf. **Buccal**). Also used in the sense of non-reproductive (structures or activities).

Spasmoneme: a contractile element in the stalk of some peritrich ciliates (Fig. 222).

Species: a taxonomic rank, representing a morphologically distinct type of organism. One definition states that individuals within a species can interbreed with each other but not with individuals from another species. This definition does not apply to many protozoa, which are not known to interbreed. Each species has two names, usually written in italics (e.g. *Elaster plaster*), the first being the name of the genus, and the second being the name of the species within the genus. The genus name has á capital initial letter, and the species name begins with a lower case letter.

Spicules: delicate, pointed structures lying external to the body, and usually directed away from it. Like spines, but invariably excreted and more delicate.

Spine: a pointed structure. Either part of a cell, or a structure secreted by a cell.

Spine scales: plate-like, secreted structures with a spine-like protrusion. Usually siliceous.

Spirotrich: a kind of ciliate (Steps 136, 161 & 178). Also called Polyhymenophora.

Stalk: a thin structure arising from the posterior part of a cell or lorica. Used for permanent or temporary attachment to the substrate. Either cytoplasmic or extracellular.

Starch grains: a term used loosely to refer to refractile masses of polysaccharides, accumulated as storage products in the cell.

Statocyst: a sensory organelle used for orientation. Typically includes a heavy weight which falls under the influence of gravity, and the direction of fall is thus sensed. Found only in one group of protists, the karyorelict ciliates.

Stigma: (plural = stigmata) organelle used in photoreception by some photosynthetic flagellates. Also called an eyespot.

Stomatocyst: a type of cyst with a siliceous wall and a single plugged opening, formed by some chrysomonads (Fig. 24).

Subapical: lying slightly away from the apex.

Subbing: a part of a culturing routine for organisms, in which a portion of an established culture is inoculated into a fresh medium.

Substrate: either the solid physical structure over which a fluid medium lies, or the material used as a basis for metabolism.

Suctoria: a type of ciliate (Step 195), with cilia only being formed in the swarmer stage. It feeds by means of projecting arms (e.g. *Podophrya*, Fig. 422).

Sulcus: a circumferential groove of some dinoflagellates. The groove carries a flagellum.

Sulphides: salts of sulphur in the reduced state, e.g. hydrogen sulphide or metal sulphides. Characteristic of reduced environments where there is no free oxygen. Sulphides cause muds to become black, usually with a smell that many non-protistologists regard as being 'bad'.

Sulphur bacteria: prokaryotes that gain energy by mediating the oxidation or reduction of sulphur compounds. Some are capable of a kind of photosynthesis, and are green or pink (purple) sulphur bacteria; others deposit elemental sulphur in their bodies. Sulphur bacteria may be abundant in some sites, being visible to the naked eye as a purple sheen on the substrate.

Suspended: said of unattached particles that float or swim in water.

Suspension feeding: feeding on suspended particles. The most usual strategy is filter-feeding, but not all suspension feeders employ this method.

Swarmer: a stage in the life cycle of some protists. Its function is distributive, i.e. it moves away from the location of the parent cell, potentially to colonize other sites. Occurs mostly as the motile stage of sessile organisms, such as peritrich ciliates (Fig. 236) or suctorian ciliates (Fig. 424).

Swimming: a form of motion in which the organism propels itself through fluid, without requiring contact with a solid substrate.

Symbiotic: living in association with another organism, normally to the mutual advantage of both, or to the advantage of one – the other being unaffected. Where there is a notable discrepancy in size, the term 'symbiont' is used to refer to the smaller member in the relationship, which may occur inside (endosymbiont) or on the outside surface (ectosymbiont) of the larger 'host' member.

Synuracea: those chrysomonads (Step 123) with a periplast of siliceous scales, spicules, etc. (e.g. *Mallomonas*, Fig. 128).

Systematics: an area of biology dealing with the classification, naming (nomenclature) and evolution of organisms.

Tardigrade: a type of very small metazoa that has an exoskeleton and eight stubby legs ending in claws.

Taxonomy: an area of biology dealing with the naming and classifying of organisms (roughly equivalent to systematics).

Test: a rigid shell around an organism, less close-fitting than a wall. Also called a lorica.

Testate: bearing a test.

Theca: a layer enclosing a body. May refer to a closely adpressed rigid wall, or to a more loosely attached rigid lorica or test, or even to a soft enclosing sheath of material.

Theront: a stage in the life cycle of some species of ciliates, in which the organism typically does not feed but moves quickly. May be thought of as an adaptation in response to a lack of food, the task of which is to hunt out new sources of food.

Trichocyst: a type of extrusome which, when extruded, takes the form of a fine, stiff filament. Typical of *Paramecium*, but used inadvertently to refer to other types of extrusomes.

Trophic: said of organisms that are active and feeding. Contrasts with the encysted state, theronts and swarmers. May also be used to refer to those aspects of metabolism associated with growth

Trophont: the feeding stage of an organism. For heterotrophic protists, this stage comprises most of the life cycle. Alternative stages are theronts (rare), swarmers and cysts.

Unequal: said of flagella of differing length on one organism. Also said of flagella which beat differently.

Underloaded: used in reference to sewage treatment works that have less incoming sewage than the plant is designed for (cf. **Overloaded**).

Undulipodia: a term preferred by some biologists, especially in the US, for eukaryotic cilia and flagella.

Undulating membrane: a part of the buccal ciliature of oligohymenophoran (and some polyhymenophoran) ciliates. A line of cilia to the left of the mouth, used to intercept particles being carried in currents of water. The particles are segregated, and passed to the cytostome for inclusion in food vacuoles. Some structures of parasitic flagellates are also referred to as undulating membranes.

Vacuole: a cavity in a cell enclosed by a membrane, e.g. food vacuoles (associated with the digestion of food) and contractile vacuoles (associated with the excretion of fluid). Small vacuoles may be called vesicles.

Ventral: relating to one side of the body. If the cell is flattened and has most of its locomotor organelles on a surface that is directed towards the substrate when moving, then that is the ventral surface. If the cell is rounded, but the mouth lies away from the apex, then the surface with the mouth is usually referred to as being ventral.

Viable: able to live.

Volvocales: those green algae (plastids with chlorophyll b and a cell wall incorporating cellulose) that normally have flagella and swim around.

Wall: a rigid layer that completely encloses a cell, and presses against the cell membrane.

Walk: a form of motion whereby the organism moves across the substrate, propelling itself with a to-and-fro movement of cilia or cirri.

Watch glass: a small, concave dish which may be used to contain fluids, cultures, etc., for observation under a binocular microscope. Solid watch glasses are squat parallelepipeds with a concave curving of one surface; they are more stable (and therefore more satisfactory) than watch glasses that are simply a thin, curved sheet of glass.

Water column: the part of a lake or pond lying between the benthos and the water–air interface.

Whiplash: a type of flagellar motion found only in some euglenids, in which a tight loop progresses from the base of the flagellum to the tip.

Xenosome: a foreign body. Used to refer both to bits of debris (which may be incorporated into the shells of some testate amoebae), and also to symbiotic algae lying inside other cells. Probably best restricted to the first usage.

Zoochlorella: endosymbiotic green algae.

Zoogloea: a slime produced by bacteria.

Bibliography

General protozoology texts, classification and techniques

A catalogue of laboratory strains of free-living and parasitic protozoa. (1958). *Journal of Protozoology,* **5**: 1–38.

de Puytorac, P., Grain, J. and Mignot, J.-P. (1987). *Précis de Protistologie.* Société Nouvelles des Editions Boubée, Paris.

Edmondson, W.T. (ed). (1959). *Ward and Whipple's Fresh-water Biology.* John Wiley & Sons Inc., New York.

Fenchel, T. (1987). *Ecology of Protozoa.* Science Tech. Publishers, Madison, Wisconsin; Springer Verlag, Berlin.

Finlay, B.J. and Ochsenbein-Gattlen, C. (1982). *Ecology of free-living protozoa.* Freshwater Biological Association, Ambleside, Cumbria.

Finlay, B.J., Rogerson, A. and Cowling, A.J. (1988). *A Beginner's Guide to the Collection, Isolation, Cultivation, and Identification of Freshwater Protozoa.* Culture Collection of Algae and Protozoa, Cumbria.

Grassé, P.P. (1952). *Traité de Zoologie:* Tome 1, Fascicule 1. Masson and Cie., Paris.

Grassé, P.P. (1953). *Traité de Zooologie:* Tome 1, Fascicule II. Masson and Cie., Paris.

Grell, K. (1973). *Protozoology.* Springer Verlag, Heidelberg.

Hausmann, K. and Patterson, D.J. (1983). *Taschenatlas der Einzeller.* Kosmos Verlag, Stuttgart.

Hausmann, K., Mulisch, M. and Patterson, D.J. (1985). *Protozoologie.* Thieme Verlag, Stuttgart.

Jahn, T.L., Bovee, E.C. and Jahn, F.F. (1979). *How to know the protozoa.* (2nd edn.), Wm. C. Brown, Dubuque, Iowa.

Kirby, H. (1950). *Materials and methods in the study of protozoa.* University of California Press, Berkeley and Los Angeles.

Kudo, R.R. (1966). *Protozoology.* (5th edn.) Charles C. Thomas Publishers, Springfield, Illinois.

Lee, J.J., Hutner, S.H. and Bovee, E.C. (1985). *An Illustrated Guide to the Protozoa.* Society of Protozoologists, Lawrence, Kansas.

Levine, N.D., Corliss, J.O., Cox, F.E.G. *et al.* (1980). A newly revised classification of the protozoa. *Journal of Protozoology,* **27**: 37–58.

Margulis, L., Corliss, J.O., Melkonian, M. and Chapman, D.J. (1990). *Handbook of Protoctista.* Jones and Bartlett, Boston, Massachusetts.

Müller, H. and Saake, E. (1979). *Mikroorganismen limnischer-Ökosysteme.* Dortmund.

Page, F.C. (1981). *The Culture and Use of Free-Living Protozoa in Teaching.* Institute of Terrestrial Ecology, Cambridge.

Pennak, R.W. (1989). *Fresh-water Invertebrates of the United States.* (3rd edn.) John Wiley & Sons Inc., New York.

Sims, R.W. , Freeman, P. and Hawksworth, D.L. (1988). *Key Works to the Flora and Fauna of the British Isles and Northwestern Europe.* (5th edn.) Oxford University Press, Oxford.

Sleigh, M.A. (1989). *Protozoa and Other Protists.* Arnold, London.

Sleigh, M.A., Dodge, J.D. and Patterson, D.J. (1984). Kingdom Protista. In *A Synoptic Classification of Living Organisms,* (ed.) R.S.K. Barnes. Blackwell, Oxford, pp.25–88.

Streble, H. and Krauter, D. (1976). *Das Leben in Wassertropfen.* Frankch'sche Verlag, Stuttgart.

Specialist literature

Allen, R.D. (1973). Contractility and its control in peritrich ciliates. *Journal of Protozoology,* **20**: 25–36.

Ammermann, D. and Schlegel, M. (1983). Characterization of two sibling species of the genus *Stylonychia* (Ciliata, Hypotricha): *S. mytilus* Ehrenberg, 1838 and *S. lemnae* n sp.I. Morphology and reproductive behaviour. *Journal of Protozoology,* **30**: 290–294.

Amos, W.B. (1972). Structure and coiling of the stalk of the peritrich ciliates *Vorticella* and *Carchesium. Journal of Cell Science,* **10**: 95–122.

Andersen, R.A. (1985). The flagellar apparatus of the golden alga *Synura uvella:* four absolute configurations. *Protoplasma,* **128**: 94–106.

Andersen, R.A. (1986). Some new observations on *Saccochrysis piriformis* Korsh. emend. Andersen (Chrysophyceae). In *Chrysophytes: Aspects and Problems,* (eds.) J. Kristiansen and R.A. Andersen. Cambridge University Press, Cambridge, pp.107–118.

Andersen, R.A. (1989). The Synurophyceae and their relationship to other golden algae. *Beiheft zur Nova Hedwigia,* **95**: 1–26.

Andersen, R.A. (1990). Three-dimensional structure of the flagellar apparatus of *Chrysopshaerella brevispina* (Chrysophyceae) as viewed by high voltage electron microscopy stero pairs. *Phycologia,* **29**: 86–97.

Anderson, O.R. and Hoeffler, W.K. (1979). Fine structure of a marine proteomyxid and cytochemical changes during encystment. *Journal of Ultrastructure Research,* **66**: 276–287.

Arnold, Z.M. (1972). Observations on the biology of the protozoan *Gromia oviformis* Dujardin. *University of California Publications in Zoology,* **100**: 1–168.

Amos, W.B. (1972). Structure and coiling of the stalk in the peritrich ciliates *Vorticella* and *Carchesium. Journal of Cell Science,* **10**: 95–122.

Balamuth, W., Bradbury, P.C. and Schuster, F.L. (1983). Ultrastructure of the amoeboflagellate *Tetramitus rostratus. Journal of Protozoology,* **30**: 445–455.

Baldock, B.M., Rogerson, A. and Berger, J. (1983). A new species of freshwater amoeba: *Amoeba algonquinensis* n. sp.(Gymnamoebia: Amoebidae). *Transactions of the American Microscopical Society,* **102**: 113–121.

Barber, H.G. and Haworth, E.Y. (1981). *A guide to the morphology of the diatom frustule.* Freshwater Biological Association, Ambleside, Cumbria.

Bardele, C.F. (1968). *Acineta tuberosa.* I. Der Feinbau des adulten Suktors. *Archiv für Protistenkunde,* **110**: 403–421.

Bardele, C.F. (1970). Budding and metamorphosis in *Acineta tuberosa.* An electron microscopic study on morphogenesis in Suctoria. *Journal of Protozoology,* **17**: 51–70.

Bardele, C.F. (1972). Cell cycle, morphogenesis, and ultrastructure in the pseudoheliozoan *Clathrulina elegans. Zeitschrift für Zellforschung,* **130**: 219–242.

Bardele, C.F. (1975). The fine structure of the centrohelidian heliozoan *Heterophrys marina. Cell and Tissue Research,* **161**: 85–102.

Bardele, C.F. (1977). Organization and control of microtubule pattern in centrohelidian heliozoa. *Journal of Protozoology,* **24**: 9–14.

Bardele, C.F. (1983). Mapping of highly ordered membrane domains in the plasma membrane of the ciliate *Cyclidium glaucoma. Journal of Cell Science,* **61**: 1–30.

Bark, A.W. (1973). A study of the genus *Cochliopodium* Hertwig and Lesser 1874. *Protistologica,* **9**: 119–138.

Batson, B.S. (1983). *Tetrahymena dimorpha* sp. nov. (Hymenostomatida: Tetrahymenidae), a new ciliate parasite of Simuliidae (Diptera) with potential as a model for the study of ciliate morphogenesis. *Philosophical Transactions of the Royal Society of London, B,* **301**: 345–363.

Belcher, J.H. and Swale, E.M.F. (1972). The morphology and fine structure of the colourless colonial flagellate *Anthophysa vegetans* (O.F. Müller) Stein. *British Phycological Journal,* **7**: 335–346.

Belcher, H. and Swale, E.M.F. (1976). *A Beginner's Guide to Freshwater Algae.* H.M.S.O., London.

Belcher, H. and Swale, E.M.F. (1979). *An Illustrated Guide to River Phytoplankton.* H.M.S.O., London.

Bellinger, E.G. (1980). *A key to common British Algae.* Institute of Water Engineers and Scientists, London.

Berger, J. and Thompson, J.C. (1960). A redescription of *Cyclidium glaucoma* O.F.M., 1786 (Ciliata: Hymenostomatida), with particular attention to the buccal apparatus. *Journal of Protozoology,* **7**: 256–262.

Bernatzky, G., Foissner, W. and Schubert, G. (1981). Rasterelektronen mikroskopische und biometrische Untersuchungen über die Variabilität der Form, Struktur und Grösse des Gehäuses einiger limnischer Tintinnina (Protozoa, Ciliophora). *Zoologica Scripta,* **10**: 81–90.

Bick, H. (1972). *Ciliated Protozoa.* World Health Organization, Geneva.

Bloodgood, R.A. (1981). Flagella-dependent gliding motility in *Chlamydomonas. Protoplasma,* **106**: 183–192.

Bohatier, J. (1970). Structure et ultrastructure de *Lacrymaria olor* (O.F.M. 1786). *Protistologica,* **6**: 331–342.

Bohatier, J. and Njiné, T. (1973). Observations ultrastructurales sur le cilié holotriche gymnostome *Litonotus quadrinucleatus* Dragesco et Njiné, 1971. *Protistologica,* **9**: 359–372.

Bonnet, L., Brabet, J., Comoy, N. and Guitard, J. (1981a). Nouvelles données sur le thécamoebien filosia *Amphitrema flavum* (Archer, 1877) Penard 1902. *Protistologica,* **18**: 225–233.

Bonnet, L., Brabet, J., Comoy, N. and Guitard, J. (1981b). Observations sur l'ultrastructure de *Nebela marginata* (Rhizopoda, Testacea, Lobosia, Hyalospheniidae). *Protistologica,* **18**: 235–241.

Borror, A.C. (1972). Revision of the order Hypotrichida (Ciliophora, Protozoa). *Journal of Protozoology,* **19**: 1–23.

Borror, A.C. (1979). Redefinition of the Urostylidae (Ciliophora, Hypotrichida) on the basis of morphogenetic characters. *Journal of Protozoology,* **26**: 544–550.

Boucaud-Camou, E. (1966). Les choanoflagellés des côtes de la Manche: I. Systématique. *Bulletin Société Linnéenne de Normandie,* Serie 10: 191–209.

Bourrelly, P. (1968). *Les algues d'eau douce. II. Les algues jaunes et brunes.* Éditions Nouvelles de Boubée et Cie, Paris.

Bourrelly, P. (1972). *Les algues d'eau douce. I. Les Algues Vertes.* Éditions Nouvelles de Boubée et Cie, Paris.

Bourrelly, P. (1985). *Les algues d'eau douce. III. Les algues bleues et rouges.* (2nd edn.) Éditions Nouvelles de Boubée et Cie, Paris.

Bourrelly, P. and Couté, A. (1981). Ultrastructure de la cuticule de quelques eugleniens: II. *Phacus horridus* Pochmann. *Protistologica,* **17**: 359–363.

Bovee, E.C. (1985a). Class Lobosea Carpenter, 1861. In *Illustrated Guide to the Protozoa,* (eds.) J.J. Lee, S.H. Hutner and E.C. Bovee. Society of Protozoologists, Kansas, pp.158–211.

Bovee, E.C. (1985b). *Class Filosea* Leidy, 1879. In *Illustrated Guide to the Protozoa,* (eds.) J.J. Lee, S.H. Hutner and E.C. Bovee. Society of Protozoologists, Kansas, pp.228–245.

Bovee, E.C. (1985c). *Order Athalamida* Haecker, 1862. In *Illustrated Guide to the Protozoa,* (eds.) J.J. Lee, S.H. Hutner and E.C. Bovee. Society of Protozoologists, Kansas, pp.246–252.

Broers, C.A.M., Stumm, C.K., Vogels, G.D. and Brugerolle, G. (1990). *Psalteriomonas lanterna* gen. nov., sp. nov., a free-living amoeboflagellate isolated from freshwater anaerobic sediments. *European Journal of Protistology,* **25**: 369–380.

Brooker, B.E. (1971). Fine structure of *Bodo saltans* and *Bodo caudatus* (Zoomastigophora: Protozoa) and their affinities with the Trypanosomatidae. *Bulletin of the British Museum (Natural History) Zoology*, **22**: 89–102.

Brugerolle, G. (1985a). Ultrastructure *d'Hedriocystis pellucida* (Heliozoa Desmothoracida) et de sa forme migratrice flagellée. *Protistologica*, **21**: 259–265.

Brugerolle, G. (1985b). Des trichocystes chez les bodonides, un caractère phylogénétique supplémentaire entre Kinetoplastida et Euglenida. *Protistologica*, **21**: 339–348.

Brugerolle, G. and Bricheux, G. (1984). Actin microfilaments are involved in scale formation of the chrysomonad cell *Synura*. *Protoplasma*, **123**: 203–212.

Brugerolle, G., Lom, J., Nohynkovà, E. and Joyon, L. (1979). Comparaison et évolution des structures cellulaires chez plusiers espèces de Bodonidés et Cryptobiidés appartenant aux genres *Bodo*, *Cryptobia* et *Trypanoplasma* (Kinetoplastida, Mastigophora). *Protistologica*, **15**: 197–221.

Burzell, L.A. (1973). Observations on the proboscis–cytopharynx complex and flagella of *Rhyncomonas metabolita* Pshenin, 1964 (Zoomastigophorea: Bodonidae). *Journal of Protozoology*, **20**: 385–393.

Buetow, D.E. (1982). *The Biology of* Euglena: Volume 3. Academic Press, New York.

Bunting, M. (1926). Studies on the life-cycle of *Tetramitus rostratus* Perty. *Journal of Morphology and Physiology*, **42**: 23–81.

Bunting, M. and Wenrich, D.H. (1929). Binary fission in the amoeboid and flagellate phases of *Tetramitus rostratus* (Protozoa). *Journal of Morphology and Physiology*, **47**: 37–87.

Butcher, R.W. (1967). *An Introductory Account of the Smaller Algae of British Coastal Waters Part IV: Cryptophyceae*. Ministry of Agriculture, Fisheries and Food, Fishery Investigations Series IV. H.M.S.O., London.

Cachon, J. and Cachon, M. (1982). Actinopoda. In *Synopsis and Classification of Living Organisms*, (ed.) S.P. Parker. McGraw-Hill, New York, pp. 553–568.

Calaway, W.T. and Lackey, J.B. (1962). Waste treatment protozoa–flagellata. Florida Engineering series, No. 3.

Cann, J.P. (1981). An ultrastructural study of *Mayorella viridis* (Leidy) (Amoebida: Paramoebidae), a rhizopod containing zoochlorellae. *Archiv. f. Protistenk.*, **124**: 353–360.

Cann, J.P. (1984). The ultrastructure of *Rhizamoeba flabellata* (Goodey) comb. nov., and *Leptomyxa reticulata* Goodey (Acarpomyxea: Leptomyxida). *Archiv. f. Protistenk.*, **128**: 13–23.

Cann, J.P. and Page, F.C. (1979). *Nucleosphaerium tuckeri* gen. nov., sp. nov. – a new freshwater filose amoeba without motile form in a new family Nucleariidae (Filosea: Aconchulinida) feeding by ingestion only. *Archiv für Protistenkunde*, **122**: 226–240.

Cash, J. and Hopkinson, J. (1905, 1909); and Cash, J., Wailes, G.H. and Hopkinson, J. (1915, 1919, 1921). *The British Freshwater Rhizopoda and Heliozoa*: Volumes 1–5. The Ray Society, London.

Chakraborty, S. and Old, K. (1986). Mycophagous soil amoebae: their biology and significance in the ecology of soil-borne plant pathogens. *Progress in Protistology*, **1**: 163–194.

Chakraborty, S. and Pussard, M. (1985). *Ripidomyxa australiensis* gen. nov., sp. nov. A mycophagous amoeba from Australian soil. *Protistologica*, **21**: 133–140.

Christopher, M. and Patterson, D.J. (1983). *Coleps hirtus*, a ciliate illustrating facultative symbioses between protozoa and algae. *Annales de la Station Biologique de Besse-en-Chandesse*, **17**: 278–296.

Collin, B. (1912). Étude monographique sur les acinetiens II. Morphologie, Physiologie, Systématique. *Archives de Zoologie expérimentale et générale*, **51**: 1–457.

Corbet, S.A. (1973). An illustrated introduction to the testate rhizopods in Sphagnum, with special reference to the area around Malham Tarn, Yorkshire. *Field Studies*, **3**: 801–838.

Corliss, J.O. (1971). Establishment of a new family (Glaucomidae n. fam.) in the holotrich hymenostome ciliate suborder Tetrahymenina, and description of a new genus (*Epenardia* n.g.) and a new species (*Glaucoma dragescui* n. sp.) contained therein. *Transactions of the American Microscopical Society*, **90**: 344–362.

Corliss, J.O. (1979). *The Ciliated Protozoa*. (2nd edn.) Pergamon, Oxford.

Corliss, J.O. and Daggett, P.M. (1983). '*Paramecium aurelia*' and '*Tetrahymena pyriformis*': current status of the taxonomy and nomenclature of these popularly known and widely used ciliates. *Protistologica*, **19**: 307–322.

Corliss, J.O. and Esser, S.C. (1984). Comments on the role of the cyst in the life cycle and survival of free-living protozoa. *Transactions of the American Microscopical Society*, **93**: 578–593.

Couté, A. and Iltis, A. (1981). Ultrastructure stéréoscopique de la logette de *Trachelomonas* (Algae, Euglenophyta) recoltés en Côte d'Ivoire. *Revue Hydrobiologie Tropicale*, **14**: 115–133.

Croome, R.L. (1986). Observations of the heliozoan genera *Acanthocystis* and *Raphidocystis* from Australia. *Archiv für Protistenkunde*, **131**: 189–199.

Croome, R.L. (1987a). *Pinaciophora columna* n. sp. and *P. apora* n. sp., new heliozoeans from Australia, and a report of *P. fluviatilis* from Antarctica. *Archiv für Protistenkunde*, **133**: 15–20.

Croome, R.L. (1987b). Observations of the heliozoea genera *Acanthocystis*, *Pompholyxophrys*, *Raphidiophrys* and *Clathrulina* from Australian freshwaters. *Archiv für Protistenkunde*, **133**: 237–243.

Curds, C.R. (1970). *An illustrated key to the British Freshwater Ciliated Protozoa commonly found in activated sludge*. H.M.S.O., London.

Curds, C.R. (1975). A guide to the species of the genus *Euplotes* (Hypotrichida, Ciliatea). *Bulletin of the British Museum (Natural History) Zoology*, **28**: 1–61.

Curds, C.R. (1982). *British and other freshwater ciliated protozoa. Part 1. Ciliophora: Kinetofragminophora.* (Synopses of the British Fauna, No. 22). Cambridge University Press, Cambridge.

Curds, C.R. (1985a). A revision of the Suctoria (Ciliophora, Kinetofragminophora) 1. *Acineta* and its morphological relatives. *Bulletin of the British Museum (Natural History) Zoology,* **48**: 75–129.

Curds, C.R. (1985b). A revision of the Suctoria (Ciliophora, Kinetofragminophora) 2. An addendum to *Acineta. Bulletin of the British Museum (Natural History) Zoology,* **49**: 163–165.

Curds, C.R. (1985c). A revision of the Suctoria (Ciliophora, Kinetofragminophora) 3. *Tokophrya* and its morphological relatives. *Bulletin of the British Museum (Natural History) Zoology,* **49**: 167–193.

Curds, C.R. (1986). A revision of the Suctoria (Ciliophora, Kinetofragminophora) 4. *Podophrya* and its morphological relatives. *Bulletin of the British Museum (Natural History) Zoology,* **50**: 59–91.

Curds, C.R., Gates, M.A. and Roberts, D.McL. (1983). *British and other freshwater ciliated protozoa. Part II. Ciliophora: Oligohymenophora and Polyhymenophora.* (Synopses of the British Fauna, No. 23). Cambridge University Press, Cambridge.

Curds, C.R. and Hawkes, H.A. (1975). *Ecological Aspects of Used-water Treatment, Volume 1 – The Organisms and their Ecology.* Academic Press, London.

Daniels, E.W. (1973). Ultrastructure. In *The Biology of Amoeba,* (ed.) K.W. Jeon. Academic Press, New York, pp.125–169.

Decloitre, L. (1976a). Le genre *Euglypha. Archiv für Protistenkunde,* **118**: 18–33.

Decloitre, L. (1976b). Le genre *Arcella* Ehrenberg. *Archiv für Protistenkunde,* **118**: 291–309.

Decloitre, L. (1977). Le genre *Nebela. Archiv für Protistenkunde,* **119**: 325–352.

Decloitre, L. (1978). Le genre *Centropyxis* I. *Archiv für Protistenkunde,* **120**: 63–85

Decloitre, L. (1979). Mises à jour au 31.12.1978 des mises à jour au 31.12.1974 concernant les genres *Arcella, Centropyxis, Cyclopyxis, Euglypha* et *Nebela. Archiv für Protistenkunde,* **122**: 387–397.

Decloitre, L. (1981). Le genre *Trinema* Dujardin, 1841. *Archiv für Protistenkunde,* **124**: 193–218.

Decloitre, L. (1982). Compléments aux publications précédentes mise à jour au 31.12.1981 des genres *Arcella, Centropyxis, Cyclopyxis, Euglypha, Nebela* and *Trinema. Archiv für Protistenkunde,* **126**: 393–407.

de Jonckheere, J.F. (1983). Isoenzyme and total protein analysis by agarose isoelectric focusing, and taxonomy of the genus *Acanthamoeba. Journal of Protozoology,* **30**: 701–706.

de la Cruz, V.F. and Gittleson, S.M. (1981). The genus *Polytomella:* A review of classification, morphology, life cycle, metabolism, and motility. *Archiv für Protistenkunde,* **124**: 1–28.

de Puytorac, P., Didier, P., Detcheva, R. and Grolière, C. (1974). Sur la morphogenèse de bipartition et l'ultrastructure du cilié *Cinetochilum margaritaceum* Perty. *Protistologica,* **10**: 223–238.

de Puytorac, P. and Njiné, T. (1970). Sur l'ultrastructure des *Loxodes* (Ciliés Holotriches). *Protistologica,* **6**: 427–444.

de Puytorac, P. and Rodrigues de Santa Rosa, M. (1975). Observations cytologiques sur le cilié gymnostome *Loxophyllum meleagris* Duj., 1841. *Protistologica,* **11**: 379–390.

de Puytorac, P. and Rodrigues de Santa Rosa, M. (1976). A propos de l'ultrastructure corticale du cilié hypotriche *Stylonychia mytilus* Ehrbg., 1838; les caracteristiques du cortex buccal adoral et paroral des Polyhymenophora Jankowski, 1967. *Transactions of the American Microscopical Society,* **95**: 327–345.

de Saedeleer, H. (1927). Notes de Protistologie I. – Craspédomonadines: Matériel systématique. *Annales de la Société Royale de Zoologie de Belgique,* **58**: 117–147.

Deroux, G. (1970). La série 'Chlamydonellienne' chez les Chlamydodontidae (Holotriches, Cyrtophorina Fauré-Fremiet). *Protistologica,* **6**: 155–182.

Deroux, G. (1974). Quelques précisions sur *Strobilidium gyrans* Schewiakoff. *Cahiers de Biologie Marine,* **15**: 571–588.

Deroux, G. (1976a). Le plan cortical des Cyrtophorida. Unité d'expression et marges de variabilité. I. – Le cas des Plesiotrichopidae, fam. nov., dans la nouvelle systématique. *Protistologica,* **12**: 469–481.

Deroux, G. (1976b). Le plan cortical des Cyrtophorida. Unité d'expression et marges de variabilité. II. – Cyrtophorida à thigmotactisme ventral generalisé. *Protistologica,* **12**: 483–500.

Deroux, G. (1976c). Plan cortical des Cyrtophorida III. – Les structures différenciatrices chez les Dysteriina. *Protistologica,* **12**: 505–538.

Didier, P. (1970). Contribution a l'étude comparée des ultrastructures corticales et buccales des Ciliés Hyménostomes Péniculiens. *Annales de la Station Biologie de Besse-en-Chandesse,* **5**: 1–274.

Didier, P. and Detcheva, R. (1974). Observations sur le cilié *Cohnilembus verminus* (O.F. Müller, 1786): morphogenèse de bipartition et ultrastructure. *Protistologica,* **10**: 159–174.

Didier, P. and Wilbert, N. (1981). Sur un *Cyclidium glaucoma* de la région de Bonn (R.F.A.). *Archiv für Protistenkunde,* **124**: 96–102.

Dobell, C. (1913). Observations on the life-history of Cienkowski's 'Arachnula'. *Archiv für Protistenkunde,* **31**: 317–353.

Dobell, C. (1960). *Antony van Leeuwenhoek and his 'Little Animals'.* Dover, New York.

Dodge, J.D. (1985). *Atlas of Dinoflagellates.* Farrand Press, London.

Dragesco, J. (1963). Révision du genre *Dileptus,* Dujardin, 1871 (Ciliata, Holotricha). *Bulletin de Biologie de la France et Belgique,* **97**: 103–145.

Dragesco, J. (1968). Les genres *Pleuronema* Dujardin, *Schizocalyptra* gen. nov. et *Histiobalantium* Stokes (Ciliés, Holotriches, Hyménostomes). *Protistologica,* **4**: 85–106.

Dunlap, J.R., Walne, P.L. and Bentley, J. (1983). Microarchitecture and elemental spatial segregation of envelopes of *Trachelomonas lefevrei* (Euglenophyceae). *Protoplasma*, **117**: 97–106.

Dürr, G. (1979). Elektronenmikroskopische Untersuchungen am Panzer von Dinoflagellaten II. *Peridinium cinctum*. *Archiv für Protistenkunde*, **122**: 88–120.

Dürrschmidt, M. (1984). Studies on scale-bearing Chrysophyceae from the Giessen area, Federal Republic of Germany. *Nordic Journal of Botany*, **4**: 123–143.

Dürrschmidt, M. (1985). Eletcron microscopic observations on scales of species of the genus *Acanthocystis* (Centrohelidia, Heliozoa) from Chile, I. *Archiv für Protistenkunde*, **129**: 55–87.

Dykstra, M.J. and Porter, D. (1984). *Diplophrys marina*, a new scale-forming marine protist with labyrinthulid affinities. *Mycologia*, **76**: 626–632.

Edmondson, W.T. (ed.) (1959). *Fresh-water Biology*. Wiley, New York.

Eikelboom, D.H. and Buijsen, H.J.J. van. (1983). *Microscopic Sludge Investigation Manual*. (2nd edn.) TNA Research Institute for Environmental Hygiene, Delft, The Netherlands.

Elliott, A.M. (1973). *Biology of* Tetrahymena. Dowden Hutchinson and Ross, Stroudsburg, Pennsylvania.

Ellis, W.N. (1929). Recent researches on the Choanoflagellata (Craspedomonadines). *Annals de la Société Royale de Zoologie de Belgique*, **60**: 49–88.

Eperon, S. (1980). Sur la stomatogenèse et les relations phylogénétiques du cilié péritriche *Thuricola folliculata* (O.F. Müller, 1786). *Protistologica*, **16**: 549–564.

Ertl, M. (1981). Zur Taxonomie der Gattung *Proterospongia* Kent. *Archiv für Protistenkunde*, **124**: 259–266.

Ettl, H. (1983). Süßwasser-flora von Mitteleuropa (Vol 9). Chlorophyta 1. Fischer Verlag, Stuttgart.

Eyden, B.P. and Kickerman, K. (1975). Ultrastructure and vacuolar movements in the free-living diplomonad *Trepomonas agilis* Klebs. *Journal of Protozoology*, **22**: 54–66.

Fenchel, T. and Small, E.B. (1980). Structure and function of the oral cavity and its organelles in the hymenostome ciliate *Glaucoma*. *Transactions of the American Microscopical Society*, **99**: 52–60.

Fenchel, T. and Finlay, B.J. (1986). Photobehavior of the ciliated protozoon *Loxodes*: taxic, transient, and kinetic responses in the presence and absence of oxygen. *Journal of Protozoology*, **33**: 139–145.

Fenchel, T. and Patterson, D.J. (1986). *Percolomonas cosmopolitus* (Ruinen) n. gen., a new type of filter feeding flagellate from marine plankton. *Journal of the Marine Biological Association, U.K.*, **66**: 465–482.

Fernandez-Galiano, D. and Fernandez-Leborans, G. (1980). *Caenomorpha medusula* Perty, 1852 (Heterotrichida, Armophorina): nouvelles données sur la ciliature et l'infraciliature. *Protistologica*, **16**: 5–10.

Finlay B.J. and Fenchel, T. (1986). Physiological ecology of the ciliated protozoon *Loxodes*. *Freshwater Biological Association Annnual Report*, **54**: 73–96.

Fischer-Defoy, D. and Hausmann, K. (1981). Microtubules, microfilaments, and membranes in phagocytosis: structure and function of the oral apparatus of the ciliate *Climacostomum virens*. *Differentiation*, **20**: 141–151.

Fleury, A., Iftode, F., Deroux, G. and Fryd-Versavel, G. (1986). Unité et diversité chez les hypotriches (Protozoaires ciliés). III. Elements d'ultrastructure comparée chez diverse représentants du sous-ordre des Pseudohypotrichina et remarques générales. *Protistologica*, **22**: 65–87.

Foissner, W. (1977). Revision der Genera *Astylozoon* (Engelmann) und *Hastatella* (Erlanger) (Ciliata Natantina). *Protistologica*, **13**: 353–379.

Foissner, W. (1979a). Peritriche Ciliaten (Protozoa: Ciliophora) aus alpinen Kleingewässern. *Zoologische Jahrbücher (Systematik)*, **106**: 529–558.

Foissner, (1979b). Morphologie, Infraciliatur und Silberliniensystem von *Phascolodon vorticella* Stein, *Chlamydonella alpestris* nov. spec. und *Trochilia minuta* (Roux) (Ciliophora, Cyrtophorida). *Protistologica*, **15**: 557–563.

Foissner, W. (1980a). Taxonomische Studien über die Ciliaten des Grossglocknergebeites (Hohe Tauern, Osterreich). IX. Ordnungen Heterotrichida und Hypotrichida. *Berichte Natur-Med. Vereins Salzburg*, **5**: 71–117.

Foissner, W. (1980b). Colpodide Ciliaten (Protozoa: Ciliophora) aus alpinen Boden. *Zoologische Jahrbücher (Systematik)*, **107**: 391–432.

Foissner, W. (1981). Morphologie und Taxonomie einiger Heterotricher und peritrichen Ciliaten (Protozoa: Ciliophora) aus alpinen Boden. *Protistologica*, **17**: 29–43.

Foissner, W. (1982). Ökologie und Taxonomie der Hypotrichida (Protozoa: Ciliophora) einiger Österreichischer Böden). *Archiv für Protistenkunde*, **126**: 19–143.

Foissner, W. (1983). Taxonomischen Studien über die Ciliaten des Grossglocknergebietes (Hohe Tauern, Osterreich). *Annalen Naturhistorische Museum Wien*, **84**(B): 49–85.

Foissner, W. (1984a). Morphologie und Infraciliatur von *Ophrydium eutrophicum* Foissner, (1979) und *Ophrydium versatile* (O.F. Müller, 1786) (Ciliophora, Peritrichida). *Berichte Naturwissenschaftlich-Medizinischen Vereins Salzburg*, **7**: 43–54.

Foissner, W. (1984b). Taxonomie und Okologie einiger Ciliaten (Protozoa, Ciliophora) des Saprobiensystems. I: Genera *Litonotus, Amphileptus, Opisthodon*. *Hydrobiologia*, **119**: 193–208.

Foissner, W. (1984c). Infraciliatur, Silberliniensystem und Biometrie einiger neuer und wenig bekannter terristrischer, limnischer und mariner Ciliaten (Protozoa: Ciliophora) aus den Klassen Kinetofragminophora, Colpodea und Polyhymenophora. *Stapfia*, **12**: 1–165.

Foissner, W. (1985). Klassifikation und Phylogenie der Colpodea (Protozoa: Ciliophora). *Archiv für Protistenkunde*, **129**: 239–290.

Foissner, W. (1988). Taxonomie und Ökologie einiger Ciliaten (Protozoa, Ciliophora) des Saprobiensystems. II. Familie Chilodonellidae. *Hydrobiologia, 162*: 21–45.

Foissner, W. (1988). Taxonomic and nomenclatural revision of Sládacek's list of ciliates (Protozoa: Ciliophora) as indicators of water quality. *Hydrobiologia, 166*: 1–64.

Foissner, W. and Adam, H. (1983). Morphologie und Morphogenese des Bodenciliaten *Oxytricha granulifera* sp. n. (Ciliophora, Oxytrichidae). *Zoologica Scripta, 12*: 1–11.

Foissner, W., Blatterer, H. and Foissner, I. (1988). The Hemimastigophora (*Hemimastix amphikineta* nov. gen., nov. spec.), a new protistan phylum from Gondwanian soils. *European Journal of Protistology, 23*: 361–383.

Foissner, W. and Foissner, I. (1988). *Catalogus Faunae Austriae, Teil 1c: Stamm Ciliophora.* Verlag der Österreichischen Akademie der Ewissenschaften, Vienna.

Foissner, W. and Rieder, N. (1983). Licht- und Raster-elektronenmikroskopische Untersuchungen über die Infraciliatur von *Loxodes striatus* (Engelmann, 1862) und *Loxodes magnus* (Stokes, 1887) (Protozoa: Ciliophora). *Zoologischer Anzeiger, 210*: 3–13.

Foissner, W. and Schiffman, H. (1974). Vergleichende Studien an Argyrophilen Strukturen von vierzehn peritrichen Ciliaten. *Protistologica, 10*: 489–508.

Foissner, W. and Schiffman, H. (1980). Taxonomie und Phylogenie der Gattung *Colpidium* (Ciliophora, Tetrahymenidae) und Neubeschreibung von *Colpidium truncatum* (Stokes, 1885). *Naturkunde Jahrbuch Stadt Linz, 24*: 21–40.

Foissner, W., Skogstad, A. and Pratt, J.R. (1988). Morphology and infraciliature of *Trochiliopsis australis* N. Sp., *Pelagohalteria viridis* (Fromentel, 1876) N. g., N. Comb., and *Strobilidium lacustris* N. Sp. (Protozoa, Ciliophora). *Journal of Protozoology, 35*: 489–497.

Foissner, W. and Wilbert, N. (1979). Morphologie, Infraciliatur und Ökologie der limnischen Tintinnina: *Tintinnidium fluviatile* Stein, *Tintinnidium pusillum* Entz, *Tintinnopsis cylindrata* Daday und *Codonella cratera* (Leidy) (Ciliophora, Polyhymenophora). *Journal of Protozoology, 26*: 90–103.

Foissner, W. and Wilbert, N. (1981). A comparative study of the infraciliature and silverline system of the fresh-water scuticociliates *Pseudocohnilembus putrinus* (Kahl, 1928) nov. comb., *P. pusillus* (Quennerstedt, 1869) nov. comb., and the marine form *P. marinus* (Thompson, 1966). *Journal of Protozoology, 28*: 291–297.

Fryd-Versavel, G., Iftode, F. and Dragesco, J. (1975). Contribution à la connaissance de quelques ciliés gymnostomes II. Prostomiens, Pleurostomiens: Morphologie, Stomatogenèse. *Protistologica, 11*: 509–530.

Gaffal, K.P. and Schneider, J. (1980). Morphogenesis of the plastidome and the flagellar apparatus during the vegetative life cycle of the colourless flagellate *Polytoma papillatum*. *Cytobios, 27*: 43–61.

Gates, M.A. (1978). Morphometric variation in the hypotrich genus *Euplotes. Journal of Protozoology, 25*: 338–350.

Giese, A.C. (1973). Blepharisma. *The biology of a light-sensitive Protozoan.* Stanford University Press, Stanford, California.

Gocht, H. and Netzel, H. (1974). Raster-elektronen-mikroskopische Untersuchungen am Panzer von *Peridinium* (Dinoflagellata). *Archiv für Protistenkunde, 116*: 381–410.

Görtz, H.-D. (1988). *Paramecium.* Springer Verlag, Berlin.

Grain, J. (1972). Étude ultrastructurale d'*Halteria grandinella* O.F.M., (Cilié Oligotriche) et considérations phylogénétiques. *Protistologica, 8*: 179–197.

Grain, J. and Golinska, K. (1969). Structure et ultrastructure de *Dileptus cygnus* Clapare de et Lachmann, 1859, cilié holotriche gymnostome. *Protistologica, 5*: 269–291.

Grassé, P.P. (1952). *Traité de Zoologie:* Tome 1, Fascicule 1. Masson and Cie., Paris.

Grassé, P.P. (1953). *Traité de Zoologie:* Tome 1, Fascicule II. Masson and Cie., Paris.

Green, J.C., Leadbeater, B.S.C. and Diver, W.L. (1989). *Chromophyte Algae: Problems and Perspectives.* Clarendon Press, Oxford.

Grolière, C.A. and Detcheva, R. (1974). Description et stomatogenèse de *Pleuronema puytoraci* n. sp. (Ciliata, Holotricha). *Protistologica, 10*: 91–99.

Grospietsch, T. (1972). *Wechsel-tierchen (Rhizopoden).* Franck'sche Verlag, Stuttgart.

Guhl, W. (1979). Beitrag zur Systematik, Biologie und Morphologie der Epstylidae (Cilia, Peritricha). *Archiv für Protistenkunde, 121*: 417–483.

Hänel, K. (1979). Systematik und Okologie der farblosen Flagellaten des Abwassers. *Archiv f. Protistenk., 121*: 73–137.

Harris, E.H. (1989). *The* Chlamydomonas *Sourcebook.* Academic Press, San Diego, California.

Harrison, F.W., Dunkelberger, D., Watabe and Stump, A.B. (1976). The cytology of the testaceous rhizopod *Lesquereusia spiralis* (Ehrenberg) Penard. *Journal of Morphology, 150*: 343–358.

Hartley, W.G. (1979). *Hartley's Microscopy.* Senecio Publishing Co., Oxford.

Hausmann, K. (1977). Bakterien und Virusähnliche Partikel im Cytoplasma des Rhizopoden *Vampyrella lateritia. Annales de la Station Biologique de Besse-en-Chandesse, 11*: 102–108.

Hausmann, K. and Peck, R.K. (1978). Microtubules and microfilaments as major components of a phagocytic apparatus: the cytopharyngeal basket of the ciliate *Pseudomicrothorax dubius. Differentiation, 11*: 157–167.

Hedley, R.H. and Wakefield, J. St.J. (1969). Fine structure of *Gromia oviformis* (Rhizopodea: Protozoa). *Bulletin of the British Museum (Natural History) Zoology, 18*: 69–89.

Hedley, R.H. and Ogden, C.G. (1973). Biology and fine structure of *Euglypha rotunda* (Testacea: Protozoa). *Bulletin of the British Museum (Natural History) Zoology, 25*: 121–137.

Hedley, R.H. and Ogden, C.G. (1974). Observations on *Trinema lineare* Penard (Testacea: Protozoa). *Bulletin of the British Museum (Natural History) Zoology,* **26**: 187–199.

Hedley, R.H., Ogden, C.G. and Krafft, J.I. (1974). Observations on clonal cultures of *Euglypha acanthophora* and *Euglypha strigosa* (Testacea: Protozoa). *Bulletin of the British Museum (Natural History) Zoology,* **27**: 103–111.

Hedley, R.H., Ogden, C.G. and Mordan, N.J. (1977). Biology and fine structure of *Cryptodifflugia oviformis* (Rhizopodea: Protozoa). *Bulletin of the British Museum (Natural History),* **30**: 313–328.

Henk, W.G. and Paulin, J.J. (1977). Scanning electron microscopy of budding and metamorphosis in *Discophrya collini* (Root). *Journal of Protozoology,* **24**: 134–139.

Herth, W. (1979). Behaviour of the chrysoflagellate alga, *Dinobryon divergens,* during lorica formation. *Protoplasma,* **100**: 345–357.

Hibberd, D.J. (1970). Observations on the cytology and ultrastructure of *Ochromonas tuberculatus* sp. nov. (Chrysophyceae), with special reference to the discobolocysts. *British Phycological Journal,* **51**: 19–143.

Hibberd, D.J. (1971). Observations on the cytology and ultrastructure of *Chrysamoeba radians* Klebs (Chrysophyceae). *British Phycological Journal,* **6**: 207–223.

Hibberd, D.J. (1975). Observations on the ultrastructure of the choanoflagellate *Codosiga botrytis* (Ehr.) Saville-Kent, with special reference to the flagellar apparatus. *Journal of Cell Science,* **17**: 191–219.

Hibberd, D.J. (1976a). The ultrastructure and taxonomy of the Chrysophyceae and Prymnesiophyceae (Haptophyceae): a survey with some new observations on the ultrastructure of the Chrysophyceae. *Botanical Journal of the Linnaean Society,* **72**: 55–80.

Hibberd, D.J. (1976b). Observations on the ultrastructure of three new species of *Cyathobodo* Petersen and Hansen (*C. salpinx, C. intricatus* and *C. simplex*) and on the external morphology of *Pseudodendromonas vlkii. Protistologica,* **12**: 249–261.

Hibberd, D.J. (1976c). The fine structure of the colonial colourless flagellates *Rhipidodendron splendidum* Stein and *Spongomonas uvella* Stein, with special reference to the flagellar apparatus. *Journal of Protozoology,* **23**: 374–385.

Hibberd, D.J. (1983). Ultrastructure of the colonial colourless zooflagellates *Phalansterium digitatum* Stein (Phalansteriida ord. nov.) and *Spongomonas uvella* Stein (Spongomonadida ord. nov.). *Protistologica,* **19**: 523–535.

Hibberd, D.J. (1985). Observations on the ultrastructure of new species of *Pseudodendromonas* Bourrelly (*P. operculifera* and *P. insignis*) and *Cyathobodo* Petersen and Hansen (*C. peltatus* and *C. gemmatus*), (Pseudodendromonadida ord. nov.). *Archiv für Protistenkunde,* **129**: 3–11.

Hibberd, D.J. and Leedale, G.F. (1985). Chrysomonadida. In *Illustrated Guide to the Protozoa,* (eds.) J.J. Lee, S.H. Hutner and E.C. Bovee. Society of Protozoologists, Lawrence, Kansas, pp.54–70.

Hilenski, L.L. and Walne, P.L. (1985). Ultrastructure of the flagella of the colorless phagotroph *Peranema trichophorum* (Euglenophyceae). II. Flagellar roots. *Journal of Phycology,* **21**: 125–134.

Hill, B.F. and Reilly, J.A. (1976). A comparative study of three fresh-water *Euplotes* species (Ciliophora, Hypotrichida). *Transactions of the American Microscopical Society,* **95**: 492–504.

Hilliard, D.K. (1971). Notes on the occurrence and taxonomy of some planktonic chrysophytes in an Alaskan Lake, with comments on the genus *Bicoeca. Archiv für Protistenkunde,* **113**: 98–122.

Hitchen, E.T. (1974). The fine structure of the colonial kinetoplastid flagellate – *Cephalothamnium cyclopum* Stein. *Journal of Protozoology,* **21**: 221–231.

Hofmann, A.H. and Bardele, C.F. (1987). Stomatogenesis in cyrtophorid ciliates. I. *Trithigmostoma steini* (Blochmann, 1895): from somatic kineties to oral kineties. *European Journal of Protistology,* **23**: 2–17.

Hollande, A. (1942). Contribution à l'étude morphologique et cytologique des genres *Biomyxa* et *Cercobodo.* Description de deux espèces coprophiles nouvelles. *Archives de Zoologie Expérimentale et Générale,* **82(1)**: 19–128.

Hollowday, E.D. (1975). Some notes on an uncommon colonial peritrichous protozoon, *Ophrydium versatile* (O.F.M.). *Microscopy,* **32**: 503–511.

Holt, J.G. (ed.) (1984–1989). *Bergey's Manual of Systematic Bacteriology.* (4 volumes) Williams and Wilkins, Baltimore, Maryland.

Huang, B. and Pitelka, D.R. (1973). The contractile process in the ciliate *Stentor coeruleus.* I. The role of microtubules and filaments. *Journal of Cell Biology,* **57**: 704–728.

Huber-Pestalozzi, G. (1955). Euglenophyceen. In *Die Binnengewasser. Part 4,* (ed.) A. Thienemann. Schweizerbart'sche Verlag, Stuttgart.

Hulsmann, N. (1982). *Vampyrella lateritia* (Rhizopoda) ingestion von *Spirogyra* – protoplasten. *Publikationen zu Wissenschaftlichen Filmen, Biologie Sektion,* **15**: 1–14.

Huttenlauch, I. and Bardele, C.F. (1987). Light and electron microscopical observations on the stomatogenesis of the ciliate *Coleps amphacanthus* Ehrenberg, 1833. *Journal of Protozoology,* **34**: 183–192.

Iftode, F., Fryd-Versavel, G. and Lynn, D.H. (1984). New details of the oral structures of *Colpidium* and *Turaniella* and transfer of the genus *Colpidium* to the Turaniellidae (Didier, 1971) (Tetrahymenina, Hymenostomatida). *Journal of Protozoology,* **20**: 463–474.

Jankowski, A.W. (1964). Morphology and evolution of Ciliophora. III. Diagnoses and phylogenesis of 53 sapropelebionts, mainly of the order Heterotrichida. *Archiv für Protistenkunde,* **107**: 185–294.

Jankowski, A.W. (1967). The boundaries and composition of the genera *Tetrahymena* and *Colpidium. Zoologischkei Zhurnal,* **46**: 17–23.

Jankowski, A.W. (1968). Morphology and systematic status of the genus *Cinetochilum* (Ciliata, Hymenostomatida). *Zoologischkei Zhurnal,* **47**: 187–194.

Jeon, K.W. (ed.) (1973). *The Biology of Amoeba.* Academic Press, New York.

Jepps, M.W. (1934). On *Kibisidytes marinus* n.gen., n.sp., and some other rhizopod protozoa found on surface films. *Quarterly Journal of Microscopical Science,* **77**: 121–127.

John, D.T. (1982). Primary amebic meningoencephalitis and the biology of *Naegleria fowleri. Annual Reviews of Microbiology,* **36**: 101–123.

Jones, A.R. (1974). *The Ciliates.* Hutchinson, London.

Joyon, L. (1965). Compléments à la connaissance ultrastructurale des genres *Haematococcus pluvialis* Flotow et *Stephanoeca pluvialis* Cohn. *Annales de la Faculté des Sciences, Clermont-Ferrand,* **26**: 57–69.

Kaczanowska, J. (1981). Polymorphism and specificity of positioning of contractile vacuole pores in a ciliate, *Chilodonella steini. Journal of Embryology and Experimental Morphology,* **65**: 57–71.

Kahl, A. (1930–1935). Wimpertiere oder Ciliata. In *Die Tierwelt Deutschlands,* (ed.) F. Dahl. Parts 18, 21, 25 & 30. Fischer Verlag, Jena.

Kaneda, M. (1960). Phase contrast microscopy of cytoplasmic organelles in the gymnostome ciliate *Chlamydodon pedarius. Journal of Protozoology,* **7**: 306–313.

Kirk, D.L., Birchem, R. and King, N. (1986). The extracellular matrix of *Volvox:* a comparative study and proposed system of nomenclature. *Journal of Cell Science,* **80**: 207–231.

Kivic, P.A. and Walne, P.L. (1984). An evaluation of the possible phylogenetic relationship between the Euglenophyta and Kinetoplastida. *Origins of Life,* **13**: 269–288.

Koonce, M.P. and Schliwa, M. (1985). Bidirectional organelle transport can occur in cell processes that contain single microtubules. *Journal of Cell Biology,* **100**: 322–326.

Kristiansen, J. (1975). On the occurrence of the species of *Synura. Verhandlungen Internationalen Vereins Limnologie,* **21**: 1444–1448

Kristiansen, J. and Andersen, R.A. (1986). *Chrysophytes: Aspects and Problems.* Cambridge University Press, New York.

Kuhlmann, S., Patterson, D.J. and Hausmann, K. (1980). Untersuchungen zu Nahrungserwerb und Nahrungsaufnahme bei *Homalozoon vermiculare* Stokes, 1887. I. Nahrungserwerb und Feinstruktur der Oralregion. *Protistologica,* **16**: 39–55.

Kühn, A. (1921). *Morphologie der Tiere in Bildern.* 1 Heft. Protozoen; 1. Teil: Flagellaten. Borntraeger, Berlin.

Kühn, A. (1926). *Morphologie der Tiere in Bildern.* 2 Heft. Protozoen; 2. Teil: Rhizopoden. Borntraeger, Berlin.

Lang, N.J. (1963). Electron-microscopic demonstration of plastids in *Polytoma. Journal of Protozoology,* **10**: 333–339.

Larsen, H.F. and Nilsson, J.R. (1983). Is *Blepharisma hyalinum* truly unpigmented? *Journal of Protozoology,* **30**: 90–97.

Larsen, J. (1985a). Ultrastructure and taxonomy of *Actinomonas pusilla,* a heterotrophic member of the Pedinellales (Chrysophyceae). *British Phycological Journal,* **20**: 341–355.

Larsen, J. (1985b). Algal Studies of the Danish Wadden Sea. II. A taxonomic study of psammobious dinoflagellates. *Opera Botanica,* **79**: 14–37.

Larsen, J. (1987). Algal studies of the Danish Wadden Sea. V. A taxonomic study of the benthic-interstitial euglenoid flagellates. *Nordic Journal of Botany,* **7**: 589–607.

Larsen, J. and Patterson, D.J. (1990). Some flagellates (Protista) from tropical marine sediments. *Journal of Natural History,* **24**: 801–937.

Laval, M. (1971). Ultrastructure et mode de nutrition du choanoflagellé *Salpingoeca pelagica* sp. nov. comparaison avec les choanocytes des spongiares. *Protistologica,* **7**: 325–336.

Leadbeater, B.S.C. (1983). Distribution and chemistry of microfilaments in choanoflagellates, with special reference to the collar and other tentacle systems. *Protistologica,* **19**: 157–166.

Leadbeater, B.S.C. and Manton, I. (1974). Preliminary observations on the chemistry and biology of the lorica of a collared flagellate (*Stephanoeca diplocostata* Ellis). *Journal of the Marine Biological Association, U.K.,* **54**: 269–276.

Leadbeater, B.S.C. and Morton, C. (1974). A microscopical study of a marine species of *Codosiga* James-Clark (Choanoflagellata) with special reference to the ingestion of bacteria. *Biological Journal of the Linnaean Society,* **6**: 337–347.

Leedale, G.F. (1985). Euglenids Bütschli, 1884. In *Illustrated Guide to the Protozoa,* (eds.) J.J. Lee, S.H. Hutner and E.C. Bovee. Society of Protozoologists, Lawrence, Kansas, pp. 41–54.

Leidy, J. (1879). *Fresh-water Rhizopods of North America.* United States Geological Survey of the Territories, Government Printing Office, Washington.

Lemmermann, E. (1914). Pantostomatinae, Promastiginae, Distomatinae. In *Die Süsswasser-Flora Deutschlands, Österreichs und der Schweiz,* (ed.) A. Pascher. I. Fischer Verlag, Jena.

Lemmermann, E. (1914). Pantostomatinae, Protomastiginae, Distomatinae, Heft 1: Flagellate 1. In *Die Süsswasser-Flora Deutschlands,* (ed.) A. Pascher. *Österreichs und der Schweiz.* Fischer Verlag, Jena.

Lindholm, T. (1985). *Mesodinium rubrum* – a unique photosynthetic ciliate *Advances in Aquatic Microbiology,* **3**: 1–48.

Lynn, D.H. (1976a). Comparative ultrastructure and systematics of the Colpodida (Ciliophora): Structural differentiation in the cortex of *Colpoda simulans. Transactions of the American Microscopical Society,* **95**: 581–599.

Lynn, D.H. (1976b). Comparative ultrastructure and systematics of the Colpodida. An ultrastructural description of *Colpoda maupasi* Enriquez, 1908. *Canadian Journal of Zoology,* **54**: 405–420.

Lynn, D.H. (1978). Size increase and form allometry during evolution of ciliate species in the genera *Colpoda* and *Tillina* (Ciliophora: Colpodida). *BioSystems*, **10**: 201–211.

Lynn, D.H. (1980). The somatic cortical ultrastructure of *Bursaria truncatella* (Ciliophora, Colpodida). *Transactions of the American Microscopical Society*, **99**: 349–359.

Lynn, D.H. (1981). The organization and evolution of microtubular organelles in ciliated protozoa. *Biological Reviews of the Cambridge Philosophical Society*, **56**: 243–292.

Lynn, D.H. (1985). Cortical ultrastructure of *Coleps bicuspis* Noland, 1925, and the phylogeny of the class Prostomea. *BioSystems*, **18**: 387–397.

Lynn, D.H. and Didier, P. (1978). Caractéristiques ultrastructurales du cortex somatique et buccal du cilié *Colpidium campylum* (Oligohymenophora, Tetrahymenina) quant à la position systématique de *Turaniella*. *Canadian Journal of Zoology*, **56**: 2336–2343.

Lynn, D.H. and Malcolm, J.R. (1983). A multivariate study of morphometric variation in species of the ciliate genus *Colpoda* (Ciliophora: Colpodida). *Canadian Journal of Zoology*, **61**: 307–316.

Machemer, H. and Deitmer, J. (1987). From structure to behaviour: *Stylonychia* as a model system for cellular physiology. *Progress in Protistology*, **2**: 213–330.

Madoni, P. (1981). *I Protozoi Ciliati Degli Impianti Biologici di Depurazione*. Consiglio Nazionale delle Recherche, Rome.

Maeda, M. and Carey, P.G. (1985). An illustrated guide to the species of the family Strombidiidae (Oligotrichida, Ciliophora), free swimming Protozoa common in the aquatic environment. *Bulletin of the Ocean Research Institute, University of Tokyo*, **19**: 1–68.

Margulis, L. and Schwartz, K.V. (1982). *Five Kingdoms: An Illustrated Guide to the Phyla of Life on Earth*. Freeman, San Francisco.

McLaughlin, R.B. (1975). *Accessories for the light microscope*. Microscope Publications Ltd., London.

Martinez, A.J. (1985). *Free-living Amebas: Natural History, Prevention, Diagnosis, Pathology, and Treatment of Disease*. CRC Press, Florida.

Maruyama, T. (1982). Fine structure of the longitudinal flagellum in *Ceratium tripos*, a marine dinoflagellate. *Journal of Cell Science*, **58**: 109–123.

Matthes, D. (1954). Suktorienstudien I. Beitrag zur Kenntnis der Gattung *Discophrya* Lachmann. *Archiv für Protistenkunde*, **99**: 187–226.

Matthes, D. (1981a). Die Familien und Gattungen der Sauginfusorien (Suctoria). *Mikrokosmos*, **70**: 161–192.

Matthes, D. (1981b). Die Familien und Gattungen der Glockentiere (Peritricha). *Mikrokosmos*, **70**: 54–58.

Matthes, D. and Guhl, W. (1973). Sessile Ciliaten der Flusskrebse. *Protistologica*, **9**: 459–470.

Matthes, D. and Guhl, W. (1975). Zwei neue Orbopercularien: *Orbopercularia finleyi* n.sp. und *Orbopercularia corlissi* n.sp. *Archiv für Protistenkunde*, **119**: 357–360.

Matthes, D., Guhl, W. and Haider, G. (1988). *Suctoria und Urceolariidae*. Fischer Verlag, Stuttgart.

Matthes, D. and Plachter, H. (1975). Suktorien der Gattung *Discophrya* als Symporionten von *Helophorus* und *Ochthebius* und also traeger symbiontischer Bakterien. *Protistologica*, **11**: 5–14.

Mignot, J.-P. (1965). Étude ultrastructurale de (*Cyathomonas truncata*) From. (Flagellé cryptomonadine). *Journal de Microscopie*, **4**: 239–252.

Mignot, J.-P. (1966). Structure et ultrastructure de quelques euglénomonadines. *Protistologica*, **2**: 51–117.

Mignot J.-P. (1974a). Étude ultrastructurale des *Bicoeca* protistes flagellés. *Protistologica*, **10**: 543–565.

Mignot, J.-P. (1974b). Étude ultrastructurale d'un protiste flagellé incolore: *Pseudodendromonas vlkii* Bourrelly. *Protistologica*, **10**: 397–412.

Mignot, J.-P. and Brugerolle, G. (1975). Étude ultrastructurale de *Cercomonas* Dujardin (= *Cercobodo* Krassilstchick), protiste flagellé. *Protistologica*, **11**: 547–554.

Mignot, J.-P., Joyon, L. and Pringsheim, E.G. (1968). Compléments a l'étude cytologique des Cryptomonadines. *Protistologica*, **4**: 493–506.

Mignot, J.-P. and de Puytorac, P. (1968). Sur le structure et la formation du style chez l'acinetien *Discophrya piriformis*. *Comptes rendus de l'Academie des Sciences de Paris*, **266**: 593–595.

Mignot, J.-P. and Savoie, A. (1979). Observations ultrastructurales sur *Nuclearia simplex* Cienkowski (Protozoa, Rhizopoda, Filosia). *Protistologica*, **15**: 23–32.

Moestrup, Ø. and Thomsen, H.A. (1976). Fine structural studies on the flagellate genus *Bicoeca* I. – *Bicoeca maris* with particular emphasis on the flagellar apparatus. *Protistologica*, **12**: 101–120.

Nanney, D.L. (1980). *Experimental Ciliatology*. Wiley, New York.

Nanney, D.L. and McCoy, J.W. (1976). Characterization of the species of the *Tetrahymena pyriformis* complex. *Transactions of the American Microscopical Society*, **95**: 664–682.

Nauss, R.N. (1949). *Reticulomyxa filosa* gen. et sp. nov., a new primitive plasmodium. *Bulletin of the Torrey Botanical Club*, **76**: 61–173.

Netzel, H. (1972). Morphogenese des Gehäuses von *Euglypha rotunda* (Rhizopoda, Testacea). *Zeitschrift für Zellforschung*, **135**: 63–69.

Netzel, H. (1975a). Die Entstehung der hexagonalen Schalenstruktur bei der Thekamöbe *Arcella vulgaris* var. *multinucleata* (Rhizopoda, Testacea). *Archiv für Protistenkunde*, **117**: 321–357.

Netzel, H. (1975b). Morphologie und Ultrastruktur von *Centropyxis discoides* (Rhizopoda, Testacea). *Archiv für Protistenkunde,* **117**: 369–392.

Netzel, H. (1977). Die Bildung des Gehäuses bei *Difflugia oviformis* (Rhizopoda, Testacea). *Archiv für Protistenkunde,* **119**: 1–30.

Nicholls, K.H. (1983a). Little-known and new heliozoeans: the centrohelid genus *Acanthocystis*, including descriptions of nine new species. *Canadian Journal of Zoology,* **61**: 1369–1386.

Nicholls, K.H. (1983b). Little-known and new heliozoeans: *Pinaciophora triangula* Thomsen new to North America and a description of *Pinaciophora pinea* sp. nov. *Canadian Journal of Zoology,* **61**: 1387–1390.

Nicholls, K.H. and Dürrschmidt, M. (1985). Scale structure and taxonomy of some species of *Raphidocystis*, *Raphidiophrys*, and *Pompholyxophrys* (Heliozoea) including descriptions of six new taxa. *Canadian Journal of Zoology,* **63**: 1944–1961.

Nisbet, B. (1974). An ultrastructural study of the feeding apparatus of *Peranema trichophorum*. *Journal of Protozoology,* **21**: 39–48.

Novotny, R.T., Lynn, D.H. and Evans, F.R. (1977). *Colpoda spiralis* sp. n., a colpodid ciliate found inhabiting treeholes (Colpodida, Ciliophora). *Journal of Protozoology,* **24**: 364–369.

Ogden, C.G. (1979a). Comparative morphology of some pyriform species of *Difflugia* (Rhizopoda). *Archiv für Protistenkunde,* **122**: 143–153.

Ogden, C.G. (1979b). An ultrastructural study of division in *Euglypha* (Protozoa: Rhizopoda). *Protistologica,* **15**: 541–556.

Ogden, C.G. (1983). Observations on the systematics of the genus *Difflugia* in Britain (Rhizopoda, Protozoa). *Bulletin of the British Museum (Natural History) Zoology,* **44**: 1–73.

Ogden, C.G. and Hedley, R.H. (1980). *An Atlas of Freshwater Testate Amoebae.* British Museum, London.

Old, K.M. and Darbyshire, J.F. (1980). *Arachnula impatiens* Cienk. A mycophagous giant amoeba from soil. *Protistologica,* **16**: 277–287.

Olive, L.S. (1975). *The Mycetozoans.* Academic Press, New York.

Ostwald, H. (1988). Ein Reise unter den Wurzelfüßern: *Reticulomyxa filosa*. *Mikrokosmos,* **77**: 123–125

Owen, H.A., Mattox, K.R. and Stewart, K.D. (1990). Fine structure of the flagellar apparatus of *Dinobryon cylindricum* (Chrysophyceae). *Journal of Phycology,* **26**: 131–141.

Owen, H.A., Stewart, K.D. and Mattox, K.R. (1990). Fine structure of the flagellar apparatus of *Uroglena americana* (Chrysophyceae). *Journal of Phycology,* **26**: 142–149.

Page, F.C. (1967). Re-definition of the genus *Acanthamoeba* with descriptions of three species. *Journal of Protozoology,* **14**: 709–724.

Page, F.C. (1977). The genus *Thecamoeba* (Protozoa, Gymnamoebia) species distinctions, locomotive morphology, and protozoan prey. *Journal of Natural History,* **11**: 25–63.

Page, F.C. (1978). An electron-microscopical study of *Thecamoeba proteoides* (Gymnamoebia), intermediate between Thecamoebidae and Amoebidae. *Protistologica,* **14**: 77–85.

Page, F.C. (1981). *Mayorella* Schaeffer, 1926, and *Hollandella* n.g. (Gymnamoebia), distinguished by surface structure and other characters, with comparisons of three species. *Protistologica,* **17**: 543–562.

Page, F.C. (1983). *Marine Gymnamoebae.* Institute of Terrestrial Ecology, Cambridge.

Page, F.C. (1983). Three freshwater species of *Mayorella* (Amoebida) with a cuticle. *Archiv für Protistenkunde,* **127**: 201–221.

Page, F.C. (1985). The limax amoebae: comparative fine structure of the Hartmanellidae (Lobosea) and further comparisons with the Vahlkampfiidae (Heterolobosea). *Protistologica,* **21**: 361–383.

Page, F.C. (1987). The classification of 'naked' amoebae (Phylum Rhizopoda). *Archiv für Protistenkunde,* **133**: 199–217.

Page, F.C. (1988). *A New Key to Freshwater and Soil Gymnamoebae.* Freshwater Biological Association, Cumbria.

Page, F.C. and Blanton, R.L. (1985). The Heterolobosea (Sarcodina: Rhizopoda), a new class uniting the Schizopyrenida and the Acrasidae (Acrasida). *Protistologica,* **21**: 121–132.

Page, F.C. and Kalinina, L.V. (1984). *Amoeba leningradensis* n. sp. (Amoebidae): a taxonomic study incorporating morphological and physiological aspects. *Archiv für Protistenkunde,* **128**: 37–53.

Page, F.C. and Siemensma, F.J. (1991). *Nackte Rhizopoda und Heliozoea.* Fischer Verlag, Stuttgart.

Patterson, D.J. (1978). *Kahl's Keys to the Ciliates.* University of Bristol, Bristol.

Patterson, D.J. (1979). On the organization and classification of the protozoon, *Actinophrys sol* Ehrenberg 1830. *Microbios,* **26**: 165–208.

Patterson, D.J. (1981). Contractile vacuole complex behaviour as a diagnostic character for free-living amoebae. *Protistologica,* **17**: 243–248.

Patterson, D.J. (1982). Photomicrography using a dedicated electronic flash. *Microscopy,* **34**: 437–442.

Patterson, D.J. (1983). On the organization of the naked filose amoeba, *Nuclearia moebiusi* Frenzel, 1897 (Sarcodina, Filosea) and its implications. *Journal of Protozoology,* **30**: 301–307.

Patterson, D.J. (1984). The genus *Nuclearia* (Sarcodina, Filosea): species composition and characteristics of the taxa. *Archiv für Protistenkunde,* **128**: 127–139.

Patterson, D.J. (1985). On the organization and affinities of the amoeba, *Pompholyxophrys punicea* Archer, based on ultrastructural examination of individual cells from wild material. *Journal of Protozoology,* **32**: 241–246.

Patterson, D.J. (1989). Stramenopiles: chromophytes from a protistan perspective. In *Chromophyte Algae: Problems and Perspectives,* (eds.) J.C. Green, B.S.C. Leadbeater and W.L. Diver. Clarendon Press, Oxford, pp.357–379.

Patterson, D.J. (1990). *Jakoba libera* (Ruinen, 1938) a heterotrophic flagellate from deep oceanic sediments. *Journal of the Marine Biological Association, U.K.,* **70**: 381–393.

Patterson, D.J. and Dürrschmidt, M. (1988). The formation of siliceous scales by *Raphidiophrys ambigua* (Protista, Centroheliozoa). *Journal of Cell Science,* **91**: 33–39.

Patterson, D.J. and Fenchel, T. (1985). Insights into the evolution of heliozoa (Protozoa, Sarcodina) as provided by ultrastructural studies on a new species of flagellate from the genus *Pteridomonas. Biological Journal of the Linnaean Society,* **34**: 381–403.

Patterson, D.J. and Fenchel, T. (1990). *Massisteria marina* Larsen and Patterson 1990, a widespread and abundant bacterivorous protist associated with marine detritus. *Marine Ecology Progress Series,* **62**: 11–19.

Patterson, D.J. and Hausmann, K. (1981). Feeding by *Actinophrys sol* (Protista, Heliozoa): I. Light microscopy. *Microbios,* **31**: 39–55.

Patterson, D.J. and Larsen, J. (1991). *The Biology of Free-living Heterotrophic Flagellates.* Oxford University Press, Oxford.

Peck, R.K. (1978). Ultrastructure of the somatic and buccal cortex of the tetrahymenine hymenostome *Glaucoma chattoni. Journal of Protozoology,* **25**: 186–198.

Peck, R.K. (1985). Feeding behaviour in the ciliate *Pseudomicrothroax dubius* is a series of morphologically and physiologically distinct events. *Journal of Protozoology,* **32**: 492–501.

Peck, R., Pelvat, B., Bolivar, I. and de Haller, G. (1975). Light and electron microscopic observations on the heterotrich ciliate *Climacostomum virens. Journal of Protozoology,* **22**: 368–385.

Penard, E. (1902). *Faune Rhizopodique du Bassin du Léman.* Kündig, Geneva.

Penard, E. (1904). *Les Héliozoaires d'eau Douce.* Kündig, Geneva.

Pennak, R.W. (1989). *Fresh-water Invertebrates of the United States.* (3rd edn.) Wiley Interscience, New York.

Pentecost, A. (1984). *Introduction to Freshwater Algae.* Richmond Publishing Co., Richmond, Surrey.

Picken, L.E.R. (1941). On the Bicoecidae: a family of colourless flagellates. *Philosophical Transactions of the Royal Microscopical Society of London,* B**230**: 451–773.

Pickett-Heaps, J.D. (1975). *Green Algae.* Sinauer Associates, Sunderland, Massachusetts.

Pochmann, A. (1942). Synopsis der Gattung *Phacus. Archiv für Protistenkunde,* **95**: 81–252.

Pontin, R.M. (1978). *A key to British Freshwater Planktonic Rotifera.* Freshwater Biological Association, Ambleside, Cumbria.

Preisig, H.R. and Hibberd, D.J. (1982a). Ultrastructure and taxonomy of *Paraphysomonas* (Chrysophyceae) and related genera 1. *Nordic Journal of Botany,* **2**: 397–420.

Preisig, H.R. and Hibberd, D.J. (1982b). Ultrastructure and taxonomy of *Paraphysomonas* (Chrysophyceae) and related genera 2. *Nordic Journal of Botany,* **2**: 601–638.

Preisig, H.R. and Hibberd, D.J. (1983). Ultrastructure and taxonomy of *Paraphysomonas* (Chrysophyceae) and related genera 2. *Nordic Journal of Botany,* **3**: 695–723.

Prescott, G.W. (1978). *How to know the freshwater Algae.* Wm. C. Brown, Dubuque, Iowa.

Pringsheim, E.G. (1946). On iron flagellates. *Philosophical Transactions of the Royal Microscopical Society of London,* B**232**: 311–342.

Pussard, M., Alabouvette, C., Lemaitre, I. and Pons, R. (1980). Une nouvelle amibe mycophage endogée *Cashia mycophaga* n. sp. (Hartmannellidae, Amoebida). *Protistologica,* **16**: 443–451.

Pussard, M. and Pons, R. (1977). Morphologie de la paroi kystique et taxonomie du genre *Acanthamoeba* (Protozoa, Amoebida). *Protistologica,* **13**: 557–598.

Rainer, H. (1968). *Urtiere, Protozoa Wurzelfüssler, Rhizopoda Sonnentierchen, Heliozoa.* Part 56 of *Die Tierwelt Deutschlands,* (ed.) F. Dahl. Fischer Verlag, Jena.

Rees, J.J., Donaldson, D.A. and Leedale, G.F. (1980). Morphology of the scales of the freshwater heliozoan *Raphidocystis tubifera* (Heliozoa, Centrohelida) and organisation of the intact scale layer. *Protistologica,* **16**: 565–570.

Repak, A.J. and Isquith, I.R. (1974). The systematics of the genus *Spirostomum* Ehrenberg, 1838. *Acta Protozoologica,* **12**: 325–333.

Richard, M. (1989). *The Bench Sheet Monograph on Activated Sludge Microbiology.* Water Pollution Control Federation, Alexandria, Virginia.

Rieder, N. (1971). Elektronenoptische Untersuchungen an *Didinium nasutum* O.F. Müller (Ciliata, Gymnostomata) in Interphase und Teilung. *Forma et Functio,* **4**: 46–86.

Roberts, D. McL., Warren, A. and Curds, C.R. (1983). Morphometric analysis of outline shape applied to the peritrich genus *Vorticella. Systematic Zoology,* **32**: 377–388.

Roberts, K.R. (1984). Structure and significance of the cryptomonad flagellar apparatus. I. *Cryptomonas ovata* (Cryptophyta). *Journal of Phycology,* **20**: 590–599.

Roberts, K.R., Stewart, K.D. and Mattox, K.R. (1981). The flagellar apparatus of *Chilomonas paramecium* (Cryptophyceae) and its comparison with certain zooflagellates. *Journal of Phycology,* **17**: 159–167.

Rodrigues de Santa Rosa, M. and Didier, P. (1975). Remarques sur l'ultrastructure du cilié gymnostome *Monodinium balbiani* (Fabre Domergue, 1888). *Protistologica,* **9**: 469–480.

Roijackers, R.M.M. and Siemensma, F.J. (1988). A study of the cristidiscoidid amoebae, with descriptions of new species and keys to genera and species (Rhizopoda, Filosea). *Archiv für Protistenkunde,* **135**: 237–253.

Sandon, H. (1927). *The composition and distribution of the Protozoan Fauna of the Soil.* Oliver and Boyd, London.

Santore, U.J. (1984). Some aspects of taxonomy in the Cryptophyceae. *New Phytologist,* **98**: 627–646.

Santore, U.J. (1985). A cytological survey of the genus *Cryptomonas* (Cryptophyceae) with comments on its taxonomy. *Archiv für Protistenkunde,* **130**: 1–52.

Sarjeant, W.A.S. (1974). *Fossil and Living Dinoflagellates.* Academic Press, London.

Schaeffer, A.A. (1926). Taxonomy of the amebas, with descriptions of thirty-nine new marine and freshwater species. *Carnegie Institute Washington Papers Dept. Marine Biology,* **24**: 1–116.

Schneider, H. (1986). Der Wasserdarm *Spongomonas intestinum. Mikrokosmos,* **75**: 174–177.

Schnepf, E. and Deichgräber, G. (1969). Über die Feinstruktur von *Synura petersenii* unter besonderer Berücksichtigung der Morphogenese ihrer Kieselschuppen. *Protoplasma,* **68**: 85–106.

Schnepf, E., Deichgräber, G., Röderer, G. and Herth, W. (1977). The flagellar root apparatus, the microtubular system and associated organelles in the chrysophycean flagellate, *Poterioochromonas malhamensis* Peterfi (syn. *Poteriochromonas stipitata* Scherffel and *Ochromonas malhamensis* Pringsheim). *Protoplasma,* **92**: 87–107.

Schrenk, H.G. and Bardele, C.F. (1991). The fine structure of *Saprodinium dentatum* Lauterborn, 1908 as a representative of the Odontostomatida (ciliophora). *Journal of Protozoology,* **38**: 278–293.

Schuster, F.L. and Pollak, A. (1978). Ultrastructure of *Cercomonas* sp., a free-living ameboid flagellate. *Archiv für Protistenkunde,* **120**: 206–212.

Shawhan, F.M. and Jahn, T.L. (1947). A survey of the genus *Petalomonas* Stein (Protozoa: Euglenida). *Transactions of the American Microscopical Society,* **66**: 182–189.

Shigenaka, Y., Watanabe, K. and Suzaki, T. (1980). Taxonomic studies on two heliozoans, *Echinosphaerium akamae* sp. nov., and *Echinosphaerium ikachiensis* sp. nov. *Annotnes Zoologicae Japonenses,* **53**: 103–119.

Siemensma, F.J. (1981). De Nederlandse Zonnendiertjes (Actinopoda, Heliozoa). *Wetenschappelijke Mededelingen K.N.N.V.,* **149**: 1–118.

Siemensma, F.J. and Roijackers, R.M.M. (1988). A study of new and little-known acanthocystid heliozoans, and a proposed division of the genus *Acanthocystis* (Actinopoda, Heliozoea). *Archiv für Protistenkunde,* **135**: 197–212.

Sims, R.W. (1980). *Animal identification:* A reference Guide. British Museum of Natural History, London.

Sims, R.W., Freeman, P. and Hawksworth, D.L. (1988). *Key Works to the Flora and Fauna of the British Isles and Northwestern Europe.* (5th edn.) Oxford University Press, Oxford.

Singh, B.N. (1981). Nuclear division as the basis for possible phylogenic classification of the order Amoebida Kent, 1880. *Indian Journal of Parasitology,* **5**: 133–153.

Singh, B.N., Misra, R. and Sharma, A.K. (1981). Nuclear structure and nuclear division as the basis for the subdivision of the genus *Thecamoeba* Fromental, 1874. *Protistologica,* **17**: 449–464.

Siver, P.A. (1991). The genus *Mallomonas.* Kluwer Publishing Group, Dordrecht.

Skuja, H. (1939). Beitrag zur Algenflora Lettlands II. *Acta Horti botanici Universitatis latviensis,* **11/12**: 41–169.

Skuja, H. (1948). Taxonomie des Phytoplanktons einiger Seen in Uppland, Schweden. *Symbolae botanicae Upsaliensis,* **9**: 1–399.

Skuja, H. (1956). Taxonomische und Biologische Studien das Phytoplankton Schwedischer Binnengewässer. *Nova Acta Regiae Societatis Scientiarum Uppsaliensis* (4th series), **16**: 1–403.

Skuja, H. (1964). Grundzüge der Algenflora und Algenvegetation der Fjeldgegenden um Abisko in Schwedish-Lappland. *Nova Acta Regiae Societatis Scientiarum Uppsaliensis* (4th series), **18**: 1–465.

Sládacek, V.V. (1972). Four mesosaprobic communities of colourless flagellates. *Archiv für Protistenkunde,* **114**: 145–148.

Sládecek, V. (1972). System of water quality from the biological point of view. *Archiv für Hydrobiologie (Beih. Ergebn. Limnol.),* **7**: 1–218.

Slankis, T. and Gibbs, S.P. (1972). The fine structure of mitosis and cell division in the chrysophycean alga *Ochromonas danica. Journal of Phycology,* **8**: 243–256.

Sleigh, M.A. (1964). Flagellar movement of the sessile flagellates *Actinomonas, Codonosiga, Monas,* and *Poteriodendron. Quarterly Journal of Microscopical Science,* **105**: 405–414

Small, E.B. and Lynn, D.H. (1981). A new macrosystem for the Phylum Ciliophora Doflein, 1901. *BioSystems,* **14**: 387–401.

Smith, R. McK. (1986). Analyses of heliozoan interrelationships: an example of the potentials and limitations of ultrastructural approaches to the study of protistan phylogeny. *Proceedings of the Royal Society of London,* B **227**: 325–366.

Society for General Microbiology (1979). The Cyanobacteria. *Journal of General Microbiology,* **111**: 1–85.

Spector, D.L. (1984). *Dinoflagellates.* Academic Press, Orlando.

Starmach, K. (1985). Chrysophyceae und Haptophyceae, Band I. In *Süsswasserflora von Mitteleuropa,* (eds.) H. Ettl, J. Gerloff, J. and H.H. Heynig. Fischer Verlag, Stuttgart.

Starr, M.P., Stolp, H., Trüper, H.G, Balows, A. and Schlegel, H.G. (eds.) (1981). *The Prokaryotes.* (2 volumes) Springer Verlag, Berlin.

Suchard, S.J. and Goode, D. (1982). Microtubule-dependent transport of secretory granules during stalk secretion in a peritrich ciliate. *Cell Motility,* **2**: 47–71.

Surek, B. and Melkonian, M. (1980). The filose amoeba *Vampyrellidium perforans* nov. sp. (Vampyrellidae, Aconchulinida): axenic culture, feeding behaviour and host range specificity. *Archiv für Protistenkunde,* **123**: 166–191.

Swale, E.M.F. (1973). A study of the colourless flagellate *Rhyncomonas nasuta* (Stokes) Klebs. *Biological Journal of the Linnaean Society*, **5**: 255–264.

Suzaki, T. and Williamson, R.E. (1986). Ultrastructure and sliding of pellicular structures during euglenoid movement in *Astasia longa* Pringsheim (Sarcomastigophora, Euglenida). *Journal of Protozoology*, **33**: 179–184.

Tamar, H. (1968). Observations on *Halteria bifurcata* sp. n. and *Halteria grandinella*. *Acta Protozoologica*, **6**: 175–183.

Tamar, H. (1974). Further studies on *Halteria*. *Acta Protozoologica*, **13**: 177–190.

Tartar, V. (1961). *The Biology of* Stentor. Pergamon, Oxford.

Tatchell, E.C. (1981). An ultrastructural study of prey capture and ingestion in *Lacrymaria olor* (O.F.M. 1786). *Protistologica*, **17**: 59–66.

Taylor, F.J.R. (1980). On dinoflagellate evolution. *BioSystems*, **13**: 65–108.

Taylor, F.J.R. (1987). *The Biology of Dinoflagellates*. Blackwell, Oxford.

Trainor, F.R. and Cain, J.R. (1986). Famous algal genera. I. *Chlamydomonas*. In *Progress in Phycological Research*, (eds.) F.R. Round and D.J. Chapman. **4**: 81–127.

Triemer, R.E. (1985). Ultrastructural features of mitosis in *Anisonema* sp. (Euglenida). *Journal of Protozoology*, **32**: 683–690.

Triemer, R.E. and Fritz, L. (1987). Structure and operation of the feeding apparatus in a colorless euglenoid, *Entosiphon sulcatum*. *Journal of Protozoology*, **34**: 39–47.

Triemer, R.E. and Ott, D.W. (1990). Ultrastructure of *Diplonema ambulator* Larsen and Patterson (Euglenozoa) and its relationship to *Isonema*. *European Journal of Protistology*, **25**: 316–320.

Tucker, J.B. (1972). Microtubule-arms and propulsion of food particles inside a large feeding organelle in the ciliate *Phascolodon vorticella*. *Journal of Cell Science*, **10**: 883–903.

Tucker, J.B. (1978). Endocytosis and streaming along highly gelated cytoplasm alongside rows of arm-bearing microtubules in the ciliate *Nassula*. *Journal of Cell Science*, **29**: 213–232.

van Bruggen, J.J.A., Stumm, C.K., Zwart, K.B. and Vogels, G.D. (1985). Endosymbiotic methanogenic bacteria of the sapropelic amoeba *Mastigella*. *FEMS Microbiology Ecology*, **31**: 187–192.

Vickerman, K. (1976). The diversity of the Kinetoplastid flagellates. In *Biology of the Kinetoplastida*, (eds.) W.H.R. Lumsden and D.A. Evans. Volume 1, Academic Press, London, pp. 1–34.

Vickerman, K. and Preston, T.M. (1976). Comparative cell biology of the Kinetoplastid flagellates. In *Biology of the Kinetoplastida*, (eds.) W.H.R. Lumsden and D.A. Evans. Volume 1, Academic Press, London, pp. 35–130.

Villeneuve-Brachon, S. (1940). Recherches sur les ciliés hétérotriches. *Archives de Zoologie expérimentale et générale*, **82**: 1–180.

Vørs, N., Johansen, B. and Havskum, H. (1990). Electron microscopical observations on some species of *Paraphysomonas* (Chrysophyceae) from Danish lakes and ponds. *Nova Hedwigia*, **50**: 337–354.

Walker, M.H., Roberts, E.M. and Usher, M.L. (1986). The fine structure of the trophont and stages in telotroch formation in *Circolagenophrys ampulla* (Ciliophora, Peritrichida). *Journal of Protozoology*, **33**: 246–255.

Walton, L.B. (1915). The Euglenoidina of Ohio. *Bulletin of the Ohio State University*, **1**: 343–350.

Warren, A. (1982). A taxonomic revision of the genus *Platycola* (Ciliophora, Peritrichida). *Bulletin of the British Museum (Natural History) Zoology*, **43**: 95–105.

Warren, A. (1985). A redescription of the freshwater loricate ciliate, *Stentor barretti* Barrett, 1870. *Archiv für Protistenkunde*, **129**: 145–153.

Warren, A. and Carey, P.G. (1983). Lorica structure of the freshwater ciliate *Platycola decumbens* Ehrenberg, 1830 (Peritrichida, Vaginicolidae). *Protistologica*, **19**: 5–20.

Wee, J.L. (1982). Studies on the Synuraceae (Chrysophyceae) of Iowa. *Bibliotheca Phycologica*, **62**: 1–183.

Weinreb, S. (1955a). *Homalozoon vermiculare* (Stokes): 1. Morphology and reproduction. *Journal of Protozoology*, **2**: 59–66.

Weinreb, S. (1955b). *Homalozoon vermiculare* (Stokes): II. Parapharyngeal granules and trichites. *Journal of Protozoology*, **2**: 67–70.

Wessel, D. and Robinson, D.G. (1979). Studies on the contractile vacuole of *Poterioochromonas malhamensis* Peterfi. I. The structure of the alveolate vesicles. *European Journal of Cell Biology*, **19**: 60–66.

Wessenberg, H. and Antipa, G. (1968). Studies on *Didinium nasutum* I. Structure and ultrastructure. *Protistologica*, **4**: 427–447.

Wessenberg, H. and Antipa, G. (1970). Capture and ingestion of *Paramecium* by *Didinium nasutum*. *Journal of Protozoology*, **17**: 250–270.

Wesenberg-Lund, C. (1925). Contributions to the biology of *Zoothamnium geniculatum* Ayrton. *Memoires de l'academie Royale des Sciences et des Lettres de Danemark, Sciences* (8th series), **10**: 1–53.

West, L.K., Walne, P.L. and Rosowski, J.R. (1980). *Trachelomonas hispida* var. *coronata* (Euglenophyceae). I. Ultrastructure of cytoskeletal and flagellar systems. *Journal of Phycology*, **16**: 489–497.

Wichterman, R. (1986). *The Biology of* Paramecium. (2nd edn.) Plenum, New York.

Wilbert, N. and Foissner, W. (1980). Eine neuebescreibung von *Calyptotricha laniginosum* Penard, 1922 (Ciliata, Scuticociliatida). *Archiv für Protistenkunde*, **123**: 12–21.

Willey, R.L., Walne, P.L. and Kivic, P. (1988). Phagotrophy and the origins of the euglenoid flagellates. *CRC Critical Reviews in Plant Sciences*, **7**: 303–340.

Williams, D.B., Williams, B.D. and Hogan, B.K. (1981). Ultrastructure of the somatic cortex of the gymnostome ciliate *Spathidium spathula* (O.F.M.). *Journal of Protozoology,* **28**: 90–99.

Williams, N.E., Buhse, H.E. and Smith, H.G. (1984). Protein similarities in the genus *Tetrahymena* and a description of *Tetrahymena leucophrys* n. sp. *Journal of Protozoology,* **31**: 313–321.

Willumsen, N.B.S. (1982). *Chaos zoochlorellae* sp. nov. (Gymnamoebia, Amoebidae) from a Danish freshwater pond. *Journal of Natural History,* **16**: 803–813.

Winkler, R.H. and Corliss, J.O. (1965). Notes on the rarely described, green colonial protozoon *Ophryidium versatile* (O.F.M.) (Ciliophora, Peritrichida). *Transactions of the American Microscopical Society,* **84**: 127–137.

Wirnsberger, E., Foissner, W. and Adam, H. (1985). Morphological, biometric, and morphogenetic comparison of two closely related species, *Stylonychia vorax* and *S. pustulata* (Ciliophora, Oxytrichidae). *Journal of Protozoology,* **32**: 261–268.

Wirnsberger, E., Foissner, W. and Adam, H. (1986). Biometric and morphogenetic comparison of the sibling species *Stylonychia mytilus* and *S. lemnaea,* including a phylogenetic system for the oxytrichids (Ciliophora, Hypotrichida). *Archiv für Protistenkunde,* **132**: 167–185.

Wohlman, A. and Allen, R.D. (1968). Structural organization associated with pseudopod extension and contraction during cell locomotion in *Difflugia. Journal of Cell Science,* **3**: 105–114.

Wu, I.C.H. and Curds, C.R. (1979). A guide to the species of the genus *Aspidisca. Bulletin of the British Museum (Natural History) Zoology,* **36**: 1–34.

Young, J.O. (1970). British and Irish freshwater Microturbellaria: historical records, new records and a key for their identification. *Archiv für Hydrobiologie,* **67**: 210–241.

Zhukov, B.F. (1971). [A key to the free-living flagellates of the suborder Bodonina Hollande]. *Trudy Institut Biologiya Vnutrennikh Vod,* **21**: 241–284 (in Russian).

Zhukov, B.F. (1978). [A guide to the flagellates of the Order Bicosoecida Grassé et Defalndre (Zoomastigophorea, Protozoa)]. *Trudy Institut Biologiya Vnutrennikh Vod,* **35**: 3–27 (in Russian).

Zhukov, B.F. and Karpov, S.A. (1985) [*Freshwater Choanoflagellates*]. Akademy Nauk S.S.S.R., Leningrad (in Russian).

Zhukov, B.F. and Mylnikov, A.P. (1983). [Colourless flagellates. Ecology and Taxonomy]. *Zoologischkei Zhurnal,* **62**: 27–41.

Zimmermann, B., Moestrup, Ø. and Hällfors, G. (1984). Chrysophyte or heliozoon: ultrastructural studies on a cultured species of *Pseudopedinella* (Pedinellales ord. nov.), with comments on species taxonomy. *Protistologica,* **20**: 591–612.

See also the journals *Microscopy* and *Mikrokosmos* (addresses below).

Addresses

The American Microscopical Society, P.O. Box 368, Lawrence, Kansas 66044, U.S.A.

The American Type Culture Collection, 12301 Parklawn Drive, Rockville, Maryland 20852, U.S.A. (source of authenticated strains).

The Carolina Biological Supply Company, Burlington, North Carolina 27215, U.S.A. (a general biological supplier in U.S.A.).

The Culture Centre of Algae and Protozoa, The Freshwater Biological Association, The Ferry House, Ambleside, Cumbria LA22 OLP, England (source of authenticated strains and teaching materials in U.K.).

Graticules Ltd., Morley Road, Tonbridge, Kent TN9 1RN, England (for micrometer slides and eyepiece graticules).

Philip Harris Ltd., Oldmixon, Weston-super-Mare, Avon BS24 9BJ, England (a general biological supplier in U.K.).

Kosmos Verlag, Pfizerstrasse, D 7000 Stuttgart, Germany (publishers of *Mikrokosmos*).

The Quekett Microscopical Club, c/o British Museum of Natural History, Cromwell Road, London SW7 5BD, England (publishers of *Microscopy*).

John Staniar and Co. Ltd., Sherborne Street, Manchester M3 1FD, England (for fine mesh netting suitable for plankton nets).

Index

All references are to page numbers.